Cognitive Science and Technology

Series Editor
David M. W. Powers, Adelaide, SA, Australia

This series aims to publish work at the intersection of Computational Intelligence and Cognitive Science that is truly interdisciplinary and meets the standards and conventions of each of the component disciplines, whilst having the flexibility to explore new methodologies and paradigms. Artificial Intelligence was originally founded by Computer Scientists and Psychologists, and tends to have stagnated with a symbolic focus. Computational Intelligence broke away from AI to explore controversial metaphors ranging from neural models and fuzzy models, to evolutionary models and physical models, but tends to stay at the level of metaphor. Cognitive Science formed as the ability to model theories with Computers provided a unifying mechanism for the formalisation and testing of theories from linguistics, psychology and philosophy, but the disciplinary backgrounds of single discipline Cognitive Scientists tends to keep this mechanism at the level of a loose metaphor. User Centric Systems and Human Factors similarly should inform the development of physical or information systems, but too often remain in the focal domains of sociology and psychology, with the engineers and technologists lacking the human factors skills, and the social scientists lacking the technological skills. The key feature is that volumes must conform to the standards of both hard (Computing & Engineering) and social/health sciences (Linguistics, Psychology, Neurology, Philosophy, etc.). All volumes will be reviewed by experts with formal qualifications on both sides of this divide (and an understanding of and history of collaboration across the interdisciplinary nexus).

Indexed by SCOPUS

More information about this series at https://link.springer.com/bookseries/11554

Victoria Kosa · Vadim Ermolayev

Terminology Saturation

Detection, Measurement and Use

Springer

Victoria Kosa
Department of Computer Science
Zaporizhzhia National University
Zaporizhzhia, Ukraine

Vadim Ermolayev
Department of Computer Science
Zaporizhzhia National University
Zaporizhzhia, Ukraine

ISSN 2195-3988 ISSN 2195-3996 (electronic)
Cognitive Science and Technology
ISBN 978-981-16-8629-0 ISBN 978-981-16-8630-6 (eBook)
https://doi.org/10.1007/978-981-16-8630-6

This Springer imprint is published by the registered company Springer Nature Singapore Pte Ltd.
The registered company address is: 152 Beach Road, #21-01/04 Gateway East, Singapore 189721, Singapore

Preface

A method and its software implementation, that help draw correct and timely conclusions and recommendations based on the analysis of big available data, are being on demand and under intensive research and development today. This analysis becomes efficient and effective if based on the knowledge elicited from data. Eliciting knowledge from a representative dataset is a challenging task due to the volume, velocity, variety, and veracity of data that have to be taken into account in a balanced manner. In Information Retrieval, the balance is captured via the use of the precision and recall measures while query answering. If a query cannot be formulated due to the broader scope of a task, e.g. developing an ontology for a subject domain, the problem becomes more challenging. In this context, a descriptive theory of the focal domain needs to be built, that reflects the requirements of all those who query the information in this domain. A way to build such a theory to be reliable and trusted by the specialist community is to ensure the completeness of the engineered ontology.

One of the widely used approaches in building descriptive domain theories is ontology learning from texts. Within this field, the requirements for building ontologies are learnt from the set of professional documents describing the domain and written by the knowledge stakeholders in the domain. It is acknowledged in this regard that the requirements are associated with the terms describing the domain. Hence, a method is needed for extracting a representative set of domain terms.

This book is a research monograph that presents the results of our work with an aim to contribute such a method that allows automatically extracting a representative set of terms, of minimal size, and forming a terminological core sub-collection of documents, of minimal size, that contains all these terms. For that, the phenomenon of terminological saturation is introduced in Chap. 1 and further studied in the book as a successive approximation process leading to the extraction of the terminological core sub-collection for the chosen arbitrary subject domain.

In Chap. 2, an extensive literature study is undertaken to figure out the proper background elements for our work. As an outcome of the review of the related work, a vision of the approach for detecting and measuring terminological saturation is proposed and a number of research questions formulated. The answers to these research questions are given in the subsequent chapters of the book and allow transform the vision into a fully elaborated and evaluated method, supported by instrumental software.

A formal framework for the method of detecting and measuring terminological saturation is developed in Chap. 3. The Chapter begins with the preliminaries that include the definitions of the basic ingredients helping us frame out the theoretical focus. Next, it presents the research hypotheses that need to be checked with regard to our framework. These hypotheses are formulated based on some of the research questions posed in Chap. 2. The remainder of Chap. 3 gives the answers to these questions. First, the formal introduction of the terminological difference function (*thd*) and the proof of its metric properties in the space of all possible document collections for a subject domain are done. Second, the formulation and proof of the theorem about the sufficient conditions for the existence of terminological saturation are provided. Finally, the optimized method for measuring terminological saturation, using the partition of a document collection is developed.

Chapter 4 is focused on the development of the algorithmic suite that operationalizes the formal framework presented in Chap. 3. It begins with the elaboration of the processing pipeline, based on the workflow for detecting and measuring terminological saturation. The workflow is further detailed in the functional block diagram that elaborates the modular structure of our algorithmic suite and dependencies between its modules. The algorithms representing the modules are further worked out to instrument all the required steps in the workflow. Finally, the chapter provides the description of the instrumental software suit that implements the algorithms and offers the pointers to this software for public use.

The objective of Chap. 5 is to present the results of the experimental evaluation of the developed method and instrumental software. It starts with the outline of experimental objectives. The explanation of out general experimental set-up is further provided and comprises the outline of the experimental workflow and toolset; the description of the document collections used as the input data for the experiments; the presentation of the measurable aspects and measures used in the experiments; and the outline of our experimental environment. The rest of the chapter focuses on the discussion of the results of the specific experimental series for evaluating or validating different aspects of the proposed solution for terminological saturation detection and measurement.

Chapter 6 is devoted to the demonstration of the use of the developed solution in industrial and academic settings. The chapter builds on the presentation of the two use cases. The first use case exploits the solution for checking the prognosis by Gartner on the potential and perspectives of industrial adoption and uptake of the Generative Adversarial Networks technology. It is done via analysing the presence or absence of terminological saturation in the document collection composed of academic, industrial, and collaborative publications in the Generative Adversarial Networks domain. The second use case demonstrates the potential of the solution for Master students to instrument their exploratory literature research for their graduate theses. Chapter 6 also deliberates on the practical advantages and limitations of the developed solution that might be experienced by the potential technology adopters.

Finally, it outlines the prognosis on the potential scope of the developed solution transfer to the industry of scholarly publishing.

Chapter 7 concludes the book with an overview of the research findings and plans for future work.

This book is intended for academics and industrial practitioners in the fields of Natural Texts Processing, Automated Term Recognition, Ontology Learning from Texts, Ontology Engineering. The book might be of interest and value to all those who process big collections of domain-bounded professional text using terminology extraction.

Zaporizhzhia, Ukraine
October 2021

Victoria Kosa
Vadim Ermolayev

Acknowledgements

The book arose from the outcome of several years of our research on finding a way to extract representative yet compact domain-bounded collections of professional documents for further domain ontology learning from these texts. This line of research, while being an essential part of the broader OntoElect project,[1] has been performed in the Ph.D. project of the first author, guided and mentored by the second author. The first author has been funded by a Ph.D. grant from the Ministry of Education and Science of Ukraine. Some part of the presented research has been funded by the stipend awarded by the Cabinet of Ministers of Ukraine. The authors are thankful for the state support of their research coming from these two governmental bodies.

The research leading to this book has been performed in part in cooperation with the Department of Computer Science of Zaporizhzhia National University, Ontology Engineering Group of the Universidad Politécnica de Madrid. The authors are thankful to the colleagues from UPM—David Chaves-Fraga, Carlos Badenes-Olmedo, and Oscar Corcho—for their invaluable help in doing joint research and contributing to publications. Some part of this project, at an early stage, has been performed in frame of the EU FP7 Marie Curie IRSES SemData project,[2] grant agreement No PIRSES-GA-2013-612551.

The collection of full-text Springer journal papers dealing with Knowledge Management, including the DMKD collection, has been provided by Springer-Verlag GmbH for our research purposes. The authors are grateful to the company for the opportunity to use this valuable resource. The authors are also thankful to Aliaksandr Birukou for his support of this fruitful cooperation with Springer-Verlag GmbH.

The author would like to express their gratitude to all the colleagues at the Department of Computer Science of Zaporizhzhia National University, in particular to Hennadiy Dobtovolskyi and Nataliya Keberle, for their generous help in software development, advice while discussing the proceedings of activities and results, leading to this book.

---

[1] https://www.researchgate.net/project/OntoElect-a-Methodology-for-Domain-Ontology-Refinement.

[2] http://www.semdata-project.eu/.

The authors express their sincere gratitude to Eugene Yuschenko, Dmytro Naumenko, and all the other colleagues at GroupBWT LLC who helped develop some tailored software and deploy the Gartner use case for the developed solution.

The authors truly appreciate and are thankful for the opportunity of trying the developed solution in academic settings, provided by the International Master Program on Computer Science and Data Science of the Ukrainian Catholic University, directed by Oleksii Molchanovskyi.

Zaporizhzhia, Ukraine
October 2021

Victoria Kosa
Vadim Ermolayev

Contents

About the Authors

Victoria Kosa has recently completed her Ph.D. with the thesis entitled "A Method of Experimental Study of Terminological Saturation in Document Collections for Knowledge Elicitation" at the Department of Computer Science of Zaporizhzhia National University (Ukraine). She is currently a self-employed researcher involved in an industrial consulting project. Her professional interests and research activities are within Automated Terminology Extraction and Ontology Engineering.

Vadim Ermolayev is an Associate Professor of Intelligent Systems at the Department of Computer Science at Zaporizhzhia National University (Ukraine). Throughout his career, Vadim combines academic activities with different professional engagements in the industry (as a research consultant) and public international organizations (as an expert) in Knowledge and Ontology Engineering, Semantic Technologies, Intelligent Software Systems, Distributed Artificial Intelligence. A particular research topic that he focuses on in his research is capturing the dynamics and adaptability of real world in intelligent artefacts. His current research interests are within Ontology Engineering, Ontology Learning, and Text Mining.

List of Figures

List of Tables

Chapter 1
Introduction

This book presents the results of our research on extracting the representative sets of terms, of minimal size, that form the terminology covering the professional needs in a subject domain—as the first phase in the domain ontology engineering pipeline. Terminologies are extracted from the collections of professional documents, e.g. scholarly publications bound to the subject domain of interest, and written by domain experts. The problem we address is challenging and has not been fully solved that far. Indeed, the solution has to ensure that the chosen collection of documents is representative of its subject domain. Further, it has to extract the core set of terms that are sufficient to represent the professional needs of the expert community around the domain. Finally, this core set of terms should be as compact as possible.

Our idea on how to address this challenge was inspired by an approach attributed to Michelangelo Buonarroti and quoted as "The sculpture is already complete within the marble block, before I start my work. It is already there, I just have to chisel away the superfluous material" (c.f. [1]). In fact, Michelangelo suggested that, if the block is good (representative) enough to contain the complete output, the art of creation becomes the technical work of removing the burden. A missing piece of the advice, however, is the method of ensuring that the block is representative and of chiselling the burden away. In the book, we develop this method, for texts (blocks) and terminologies contained in these texts (sculptures), based on the use of the phenomenon of terminological saturation. In difference to the approach by Michelangelo, we do not start with the whole block and do not chisel the burden away. On the contrary, we discover and pick the gems in the rock, putting these to the artefact, incrementally. We stop when the artefact is saturated with the gems. As we prove in the book, a method based on this idea could indeed cope with the scalable knowledge extraction from big textual data available for many real-world subject domains.

A method and its software implementation, that help draw correct and timely conclusions and recommendations based on the analysis of big available data, through knowledge extraction, are on demand and under intensive research and development today. As pointed out by Ermolayev et al. [2], one of the challenges in this way is

V. Kosa and V. Ermolayev, *Terminology Saturation*, Cognitive Science and Technology,
https://doi.org/10.1007/978-981-16-8630-6_1

resolving the inherent contradiction between the efficiency and effectiveness of the approach. The complexity of the challenge is amplified due to several dimensions of data, volume, velocity, variety, veracity, that have to be taken into account in a balanced manner. For example, with the increase of the volume, the efficiency decreases but the effectiveness might increase. Differently from the volume, the increase in velocity decreases effectiveness.

In Information Retrieval, the above-mentioned contradiction is somewhat captured by the precision and recall measures [3]. Indeed, with the growth of data, recall might increase, as more true positives might be present in a bigger sample. However, precision might decrease as more false positives might occur in a bigger sample along with the true positives. Recall and precision are often balanced using F-measure [3]. In addition to this known balanced measurement, a valid question would be: what is the data sample of a minimal size that contains all the true positives? Answering this question enables the most efficient and scalable method of retrieving information as it allows filtering out all redundant data. Such a data sample of minimal size is, straightforwardly, representative in the context of the corresponding information retrieval query.

1.1 Representativeness Challenge in Ontology Engineering

A more complicated question arises if we broaden the scope of a query to a particular subject domain of discourse: is there a data sample of the minimal size for a subject domain, which is representative for any relevant query within the domain? The question is complicated, as we have to know all the true positives for any query about the domain. To answer this question, a descriptive theory of the domain needs to be built, that reflects the requirements of all those who query the information in this domain. A way to build such a theory is engineering a domain ontology that is sufficiently complete to be shared by the community of the knowledge stakeholders of the domain. Hence, the initial question is transformed to how to ensure that the developed domain ontology is complete.

The mainstream practice in Ontology Engineering is collecting relevant competency questions via organizing systematic interviews or brainstorming sessions with the groups of experts selected from the knowledge stakeholders (c.f. [4–8]). The answers to the competency questions are further used as the requirements for building the ontology. The problem, however, is that these methodologies do not guarantee that the selected group of experts represents the sentiment of the community satisfactorily fully. Hence, the collected requirements set cannot be regarded as representative.

A relatively more recent trend in building descriptive domain theories is ontology learning from texts [9]. Within this field of research, the requirements for building ontologies are learnt from the set of professional documents describing the domain and written by the knowledge stakeholders for the domain. It is acknowledged in this context that the requirements are associated with the terms describing the domain. Therefore, if a representative set of relevant terms is available, then the requirements for a complete ontology could be unfolded from these terms. Hence, the method is required to extract a representative set of terms describing the domain. For that, a terminologically complete collection of the documents, describing the domain, needs to be available.

In this work, such a method is proposed that allows automatically extracting a representative set of terms of minimal size and forming a terminological core sub-collection of documents, of minimal size, that contains all these terms. For that, the notion of terminological saturation as a successive approximation process leading to the terminological core sub-collection for an arbitrary subject domain is introduced.

1.2 Phenomenon of Saturation

Saturation, as denoted by Oxford Learners Dictionaries,[1] is "the state or process that happens when no more of something can be accepted or added because there is already too much of it or too many of them". The term is broadly applicable to label the phenomena in material and immaterial worlds. Some examples are:

- In Chemistry, saturation means "the degree to which something is absorbed in something else, expressed as a percentage of the greatest possible".[1]
- In Magnetism, it stands for "the state when a magnetic material is fully magnetized".[2]
- In Thermodynamics, it points to "thermodynamic state at lower temperature bound of superheated steam".[2]
- In Commutative Algebra, "saturation with respect to f of an ideal I in R is the inverse image of $R_f I$ this ideal under the canonical map from R to R_f. It is the ideal consisting in all elements of R whose products by some power of f belong to I."[3]
- In Economics, market saturation is the state of a market in which a product has become distributed within the market up to the natural consumption level [10].
- In Genetics, saturation is "the result of multiple substitutions at the same site in a sequence, or identical substitutions in different sequence, such that the apparent sequence divergence rate is lower than the actual divergence that has occurred" [11].
- In Qualitative Research, data saturation stands for "the phase of qualitative data analysis in which the researcher has continued sampling and analysing data until no new data appear and all concepts of the theory are well developed … and their linkages to other concepts are clearly described" [12].

In this work, terminological saturation is studied as a phenomenon in the sequence of incrementally growing document sub-collections of a (hypothetical) complete

[1] https://www.oxfordlearnersdictionaries.com/definition/american_english/saturation.

[2] https://en.wikipedia.org/wiki/Saturation.

[3] https://en.wikipedia.org/wiki/Gr%C3%B6bner_basis#Definition_of_the_saturation.

collection. The process of terminological saturation is modelled as a successive approximation process. It is found, experimentally, and proven, formally, that terminological saturation results in a saturated sub-collection of documents under some conditions in a domain-bounded setting. This saturated sub-collection, termed as a terminological core sub-collection, appears to be similarly representative, as the entire collection, to reflect the sentiment of the knowledge stakeholders about the domain. Therefore, the set of terms extracted from the terminological core sub-collection could be regarded as a minimal representative term-set for further requirements elicitation in the ontology learning pipeline.

1.3 Structure of the Book

Chapter 2 reviews the related work that helped us develop our approach to model terminological saturation, detect, and measure the result of this process. It starts with the outline of the methodology used to select the literature sources for our systematic literature review in Sect. 2.1. Further, in Sects. 2.2–2.11 it reviews and analyses the relevant State-of-the-Art in various research fields ranging from Ontology Engineering and Ontology Learning to Qualitative Research. The research gaps found in this analysis are summarized and our motivation to narrow these gaps is explained in Sect. 2.12. Based on the outlined research gaps, Sect. 2.13 offers a roadmap for resolving the discovered open issues. It proposes a vision of and outlines the possible approach to the solution for narrowing the discovered research gaps. Further, it formulates the research questions to be answered within the elaboration of the envisioned approach. Based on the research questions, the research objectives of the presented study are set. Section 2.14 concludes the chapter by summarizing the findings.

Chapter 3 focuses on the elaboration of the formal framework for terminological saturation measurement as a method. The chapter begins with the preliminaries (Sect. 3.1) that include the definitions helping us frame out the theoretical focus. Next, the research hypotheses are presented to be further checked with regard to our theoretical framework. These hypotheses are formulated based on the research questions (Sect. 2.13). Sections 3.3 and 3.4 focus on the formal introduction of the terminological difference function (thd) and the proof of its metric properties in the space of all possible document collections for a subject domain. Section 3.5 formulates and proves the theorem on sufficient conditions for the existence of terminological saturation. Finally, Sect. 3.6 investigates if the method for measuring and detecting terminological saturation could be optimized in terms of lowering the incurred computational overhead. As a result, the optimized method is elaborated that uses the partitioning of a document collection. It is proven that the optimized method yields the same thd measurement values as the baseline, but spends substantially less time for computation. Furthermore, it is argued that the optimized method is parallelizable. Hence, in addition to its efficiency, it waives the constraints inherent for any straightforward implementation of an incremental way of computing saturation measurements using the elaborated baseline method (c.f. Sects. 3.1–3.5).

Chapter 4 presents the development of new or improvement of formerly developed algorithms that operationalize our formal framework (Chap. 3) for computation. It begins with the elaboration of the processing pipeline in Sect. 4.1. For that, the workflow for measuring terminological saturation (Fig. 4.1) is proposed. The workflow is further detailed in the functional block diagram that reveals the modular structure of our algorithmic suite and dependencies between its modules (Fig. 4.3). The algorithms representing the modules are then developed as follows. Section 4.2 presents the algorithms developed for instrumenting data preparation. These include the generation of a document collection catalogue and the download of publicly available full-text documents. Section 4.3 presents the algorithms developed for the pre-processing phase of the workflow. These are for PDF to plain text conversion and dataset generation based on the defined configuration parameters. The optimized algorithm for the statistical part of the term extraction pipeline and the algorithm for merging partial C-Values are elaborated in Sect. 4.4. The baseline algorithm for computing terminological difference between two bags of terms is presented in Sect. 4.5. Section 4.6 refines this baseline algorithm by incorporating the developed term grouping technique, including the algorithms for string similarity measurement and the refined algorithm for measuring terminological difference. The algorithm for removing regular accumulated noise from the bags of terms is reported in Sect. 4.7. Finally, Sect. 4.8 outlines the implementation of the algorithmic suite in software and offers the links to this software, to make it publicly available for academic research use.

Chapter 5 presents our experimental evaluation of the developed method (Chap. 3) and algorithmic suite (Chap. 4) for terminological saturation measurement. It starts with formulating the experimental objectives in Sect. 5.1. It continues with the set-up of our experiments in Sect. 5.2. Sections 5.3–5.7 report and discuss the results of the evaluation experiments. As explained in the chapter, the results prove the correctness, domain neutrality, efficiency, and scalability of the developed method and algorithmic suite.

Chapter 6 describes our experience of the use and views on how the presented approach could be more effectively transferred into academic and industrial practice. Section 6.1 presents our experience of exploiting the developed software pipeline in an industrial project for cross-checking the prognosis by Gartner,[4] on the trends of technology uptake, for the Generative Adversarial Networks [13] technology of Deep Learning (e.g. [14]). Section 6.2 reports on the academic use of our method and software for instrumenting the exploratory research by Master students, with their aim to write the reviews of the related work for their Master theses. Section 6.3 extends the use case experience, reported in Sects. 6.1 and 6.2, by outlining the practical benefits and limitations of our method and software for the potential early industrial adopters. Section 6.4 presents our view on the potential business scenarios and applications in the scholarly publishing industry. A summary of the experience and prospects for the practical use of our contribution is given in Sect. 6.5.

[4] https://www.gartner.com/en.

Finally, we conclude on our findings and present the plans for future work in Chap. 7.

References

1. Frasca S, Leghziel RC, Arabadzhiev IN et al (2021) The Michelangelo step: removing scalloping and tapering effects in high aspect ratio through silicon vias. Sci Rep 11:3997. https://doi.org/10.1038/s41598-021-83546-w
2. Ermolayev V, Akerkar R, Terziyan V, Cochez M (2013) Toward evolving knowledge ecosystems for big data understanding. In: Akerkar R (ed) Big data computing. Taylor & Francis, Boca Raton, FL, pp 3–56
3. Baeza-Yates R, Ribeiro-Neto B (1999) Modern information retrieval. ACM Press, Addison-Wesley, New York, NY. ISBN: 0-201-39829-X
4. Sure Y, Staab S, Studer R (2003) On-To-Knowledge methodology. In: Staab S, Studer R (eds) Handbook on ontologies. International handbooks on information systems. Springer, Berlin, Heidelberg, pp 117–132. https://doi.org/10.1007/978-3-540-24750-0_6
5. Gómez-Pérez A, Fernández-López M, Corcho O (2004) Ontological engineering. Springer, London.https://doi.org/10.1007/b97353
6. Pinto HS, Tempich C, Staab S, Sure Y (2004). DILIGENT: towards a fine-grained methodology for distributed, loosely-controlled and evolving engineering of ontologies. In: de Mántaras RL, Saitta L (eds) ECAI'04: 16th European conference on artificial intelligence, Valencia, Spain, 2004. IOS Press, pp 393–397. https://doi.org/10.5555/3000001.3000084
7. Schreiber G, Akkermans H, Anjewierden A, de Hoog R, Shadbolt NR, Van de Velde W, Wielinga BJ (1999) Knowledge engineering and management: the CommonKADS methodology. The MIT Press, Cambridge, MA
8. Suárez-Figueroa MC, Gómez-Pérez A, Motta E, Gangemi A (2012) Ontology engineering in a networked world. Springer, Heidelberg.https://doi.org/10.1007/978-3-642-24794-1
9. Wong W, Liu W, Bennamoun M (2012) Ontology learning from text: a look back and into the future. ACM Comput Surv 44(4):20:1–20:36. https://doi.org/10.1145/2333112.2333115
10. Osenton T (2004) The death of demand: finding growth in a saturated global economy. Financial Times Prentice Hall Books, Upper Saddle River, NJ. ISBN: 9780131423312
11. Philippe H, Brinkmann H, Lavrov DV, Littlewood DT, Manuel M, Wörheide G, Baurain D (2011) Resolving difficult phylogenetic questions: why more sequences are not enough. PLoS Biol 9(3):e1000602. https://doi.org/10.1371/journal.pbio.1000602
12. Morse JM (2004) Theoretical saturation. In: Lewis-Beck MS, Bryman A, Liao TF (eds.) The Sage encyclopedia of social science research methods. Sage, Thousand Oaks, CA, p 1123. http://sk.sagepub.com/reference/download/socialscience/n1011.pdf. Accessed 14 Oct 2021
13. Goodfellow I, Pouget-Abadie J, Mirza M, Xu B, Warde-Farley D, Ozair S, Courville A, Bengio Y (2014) Generative adversarial nets. In: Ghahramani Z, Welling M, Cortes C, Lawrence ND, Weinberger KQ (eds) Advances in neural information processing systems, vol 27. Curran Associates, Inc., Boston, pp 2672–2680
14. Schmidhuber J (2015) Deep learning. Scholarpedia 10(11):32832. https://doi.org/10.4249/scholarpedia.32832

Chapter 2
Related Work and Our Approach

The problem of extracting saturated term-sets for learning domain ontologies from professional texts, describing a subject domain of interest, appeared to be under-researched. Therefore, a broader scale systematic review of the related work has been undertaken for collecting different bits of relevant knowledge about the State-of-the-Art in various fields. These ranged from Ontology Engineering, through Ontology Learning from Texts and Information Science, to Qualitative Research in Social and Medical Sciences.

2.1 Methodology for Literature Sampling

For our systematic literature review, we used the methodology and software pipeline by Dobrovolskyi and Keberle [1] to collect and ensure the completeness of the relevant set of papers for further analysis. Their approach puts together snowball sampling, probabilistic topic modelling, and citation network analysis to search through the titles, keywords, and abstracts of scientific papers indexed by Microsoft Academic.[1].

Dobrovolskyi and Keberle suggested [1] to manually selecting:

- Either a small set of very relevant and well-cited review papers on the topic of interest
- Or a set of seed key phrases that best characterized the topic of interest.

---

[1] https://academic.microsoft.com/.

V. Kosa and V. Ermolayev, *Terminology Saturation*, Cognitive Science and Technology,
https://doi.org/10.1007/978-981-16-8630-6_2

As we were not informed about a review paper on terminology saturation in automated terms recognition or ontology learning, we followed the second path and provided several relevant seed key phrases: automated term extraction; automated term recognition; theoretical saturation; terminology saturation; terminological saturation; corpus completeness; ontology learning from text; saturation measure; terminology saturation measure.

The output of literature sampling provided two disjoint clusters of papers on[2]: (i) ontology learning from texts and ATE (474 papers); and (ii) theoretical (data) saturation in Qualitative Research (512 papers). The scan of the titles and abstracts of the papers within the first cluster revealed for: "saturation"—0 papers; "completeness"—4 papers, 3 of which were marginally relevant and 1 by Ferrari et al. [2] was sufficiently relevant. In the second cluster, we found 12 papers for "saturation" and 4 papers for "completeness". We used 2 of that 12 papers mentioning "saturation" [3, 4] in our review (Sect. 2.6 Theoretical Saturation and Ontology Learning). At the second iteration, we took a broader look at literature, which involved: the search for relevant papers in the focal fields of study; and making use of the citation network based on the relevant papers found at the first iteration. The results of the analysis of the identified relevant publications are provided in subsequent sub-sections.

2.2 Domain Ontology Engineering and Requirements Elicitation

An ontology as an artefact [5], by definition, is "a formal, explicit specification of a shared conceptualization" [6]. A mainstream interpretation of this term in Ontology Engineering for domain ontology is that it is a formal theory describing the subject domain. An important property of a domain ontology, with respect to our focus of interest, is that it has to be a shared specification. A mainstream assessment of being shared is the degree to which an ontology supports the interpretations of or views on the subject domain by the domain professionals—knowledge stakeholders. The more views, further termed as requirements, are supported by the ontology, the higher is the acceptance of this ontology by the knowledge stakeholders. Therefore, it is more shared by them.

To design an ontology that adequately meets the requirements of the focal community of knowledge stakeholders, it is imperative to be informed sufficiently fully about their views. This poses a challenge, as it is hard to elicit directly from the knowledge stakeholders the interpretations of the domain in an explicit form [7]. Different ontology engineering methodologies attack this challenge in slightly different manners, all based on organizing systematic interviews or brainstorming sessions with the experts selected from the knowledge stakeholders [8–12]—to

[2] The lists of papers, including citation counts, titles, abstracts, and Microsoft Academic links jhave been made available to public at https://github.com/OntoElect/LR. Filenames: for cluster (i) Publist-OLT-474.xlsx; for cluster (ii) Publist-TS-551.xlsx.

mention the few most frequently cited. However, there is always a risk that the selected group and their requirements under-represent the sentiment of the specialist community.

Unfortunately, no guidance was found in ontology engineering literature on how to measure objectively the representativeness of an expert group and, therefore, the completeness of the requirements elicited from them. In fact, as the experts are expensive, the trade-off between the completeness and the price was typically made in favour of lowering the price.

Crowdsourcing approaches to requirements elicitation in ontology engineering [13] gained momentum as a lower-cost way "to distribute the completion of a task among a potentially large group of contributors". Despite several reports about the success stories on the use of crowdsourcing in the Semantic Web area (c.f. [13]), the questions about the completeness of the requirements elaborated by a crowd-sourced team and the lack of the bias in domain interpretations largely remained unanswered to date.

2.3 Ontology Learning from Texts and Community Consensus

To address the difficulties in ensuring representativeness and lowering bias, it has been proposed, by Maedche and Staab [14], to learn ontologies or the requirements for ontology development (also termed as features), not from the group of experts, but from the artefacts developed by knowledge stakeholders in their subject domain. The rationale was that extracting features from the available artefacts was less laborious and took account of the plurality of opinions. In addition, it could have been done using partial automation. Furthermore, collecting a representative sample of artefacts was more feasible than a representative group of professionals possessing appropriate expertise.

One of the relevant types of these artefacts is professional textual documents, of which scholarly publications represent the high-quality part. Ontology learning from texts is being, to date, a noticeable subfield in ontology learning with developed methodologies and processing pipelines [15–17]. To learn the requirements for engineering a domain ontology, a representative set of textual documents needs to be collected. This document collection has to:

- Contain relevant texts of sufficiently good quality
- Be sufficiently complete (representative) to reflect community consensus.

There is a bunch of approaches to collect relevant documents (Sect. 2.4). However, the problem of checking if a document collection is sufficiently complete has not yet been fully resolved, to the best of our knowledge.

Perhaps, a reason for under-estimating the importance of ensuring the representativeness of a text collection is that textual resources are abundant. Hence, one may always expect to be able to have enough, even for a narrowly focused subject domain. For example, it is feasible to collect high-quality research papers on a particular topic or within a field, as we did for Knowledge Management (KM) [18]. This KM collection contains over 7,400 journal articles in full texts, which might be considered as a sufficient volume due to the recommendations of linguistic corpora experts (c.f. [19]). However, even if one collects what she thinks enough, there ought to be a way to measure the representativeness of this text corpus with regard to ontology learning.

Recently, several efforts were undertaken to extract ontologies from scholarly data—which were scientific publications—in an automated way. One topical example was the extraction of the Computer Science Ontology (CSO) by Salatino et al. [20]. CSO was extracted using the Klink-2 algorithm [21]. Klink-2 appeared to be scalable enough to generate the complete CSO using the collection of the metadata of 16 million scientific papers. Using only metadata, even in the combination with external resources, like DBpedia[3] entries, allowed however only for shallow document analysis.

In our work, in difference to the Klink-2 approach, we process the full texts of scientific papers and often filter out significant parts of metadata, like venue and author names, which appear to be the parts of accumulated regular noise, as it might be observed in our experiment with the DAC Naturelle collection (c.f. Sect. 5.5.3).

2.4 Collecting Relevant Documents of Good Quality

A reliable way to collect documents of good quality is to look at peer-reviewed scholarly publications. Peer review assures both domain relevance and professional quality of a text. The challenge is that the body of scientific papers is vast. Hence, manual search for and selection of the representative collection of papers, relevant for a domain or researched topic within a domain, is not feasible.

Several published approaches for collecting relevant papers in manual [22] or (partially) automated [23–26] ways are known from the literature. The techniques used in these approaches were: constrained snowball sampling followed by citation network analysis [23, 24, 26]; exploratory search based on probabilistic topic modelling [25]; or extensive systematic experts labour [22] combining snowball sampling, keyword search, and reading.

To select relevant sources for our literature review in the work presented in this book, we used the method and instrumental pipeline by Dobrovolskyi and Keberle [1] as outlined in Sect. 2.1. The method was the combination of automated probabilistic topic modelling, snowball sampling, and citation network analysis. In some cases, as for example for the RAW document collection (Sect. 5.2.2), we also used the manual approach based on the expert judgements to select document collections used in our experiments (Sect. 5.2).

[3] https://wiki.dbpedia.org/.

2.5 Terminological Saturation and Representativeness

As revealed by our literature sampling result (Sect. 2.1), the use of the phenomenon of saturation was not sufficiently well researched in Computer or Information Science, in particular in Text Mining or Ontology Learning. Riazanov and Voronkov [27] used the saturation of clauses to optimize automated theorem proving. Chernyak and Berenstein [28] used term saturation for building term-proximity graphs in document clustering and query answering. Doerre et al. [29] exploited the phenomenon of saturation to sample properly a subset of documents for performing further hierarchical cluster analysis in building a topic taxonomy. Han et al. [30] proposed to use a saturation measure to cluster patent texts and classify these by topics.

Ferrari et al. [2] proposed two metrics, for assessing the completeness of software requirements specifications, based on the account for relevant terms and their relationships. In their work, terms and relationships have been extracted from the requirements specifications that had been available as natural language texts. The approach by Ferrari et al. [2] helped assess the completeness of the specified requirements by comparing these to what has been extracted from the available single document or a small set of documents. However, it was not helpful in finding out if the used set of documents was complete enough to represent appropriately the majority opinion within the knowledge stakeholder community. Despite that, the approach by Ferrari et al. [2] resembles our work in the use of terminology extraction for assessing completeness.

In difference to Ferrari et al. [2], our approach (c.f. Chaps. 3 and 4) allows not only measuring and detecting terminological saturation, but also extracting a terminological core sub-collection if saturation has been detected. Hence, it constructively enables retaining compact, yet representative, sets of significant terms from the corresponding terminological core sub-collections.

2.6 Theoretical Saturation and Ontology Learning

The publications in the field of Qualitative Research have been reviewed in our literature study because their objective was somewhat coherent to Ontology Learning. Indeed, the methodology of qualitative research is based on processing interviews. The objective of this processing is to build a (descriptive) theory to support (or refute) a research hypothesis in the researched context. The complication faced by qualitative analysts is also similar to that of ontology engineers (c.f. Sect. 2.2)—interviews are expensive. This is why, and in difference to ontology engineers, qualitative researchers sought a way to circumscribe a sample of subjects that already gave all the replies that would have been received from the other subjects who have not been interviewed yet.

The pioneers who introduced data saturation in Qualitative Research were Glaser and Strauss [4]. They proposed the Grounded Theory method for processing interview data and denoted this method as "the discovery of theory from data systematically obtained from social research" [4]. The groundedness of a theory, in their view, meant that every statement in that theory was supported by the collected data. Glaser and Strauss [4] claimed that a theory became grounded if it had been discovered systematically. However, they did not reveal how to operationalize this discovery for making it systematic, except giving a few heuristic suggestions.

Attempts to operationalize the detection of data saturation in the context of the Grounded Theory method have been made and broadly discussed in the Qualitative Research literature. The analysis of these proposals was, however, beyond the scope of our work. Instead, we referred to the conclusion of one of the recent publications by Aldiabat and Navenec [3]. They acknowledged that the detection of data saturation was still a "mysterious step". Further, it needs to be mentioned that a reason for this lack of progress might have been that the identification of the factors that facilitate or hinder data saturation was done informally. With respect to terminological saturation, a formal measure of it, with formally proven validity, was still not available in the literature. One early paper attempting to propose an algorithmic approach to measure terminological saturation was [7].

The Grounded Theory method in Qualitative Research suggested that saturation checks had to be performed every time a new portion of data was taken into consideration. Hence, the method was iterative by its nature. The question, not fully and rigorously answered by sociologists, was: what further data items (documents) needed to be taken into analysis to ensure that the saturated sample was as compact and stable[4] as possible? The recommendations given by Aldiabat and Navenec [3] included: accurate scoping of research question; using data collection method triangulation; using experienced researchers familiar with the research method; understanding the philosophical underpinnings of the method; using sensitizing concepts.

In difference to our work, these recommendations, however, did not offer any objective rule for ordering the input documents for inclusion into the upcoming portion of input data.

2.7 Ordering of Documents for Processing

To our regret, the reviewed related work offered insufficiently comprehensive and grounded recommendations on how to order the documents, in iterations, for extracting compact and stably saturated sets of terms to be further used as the features for systematic requirements elicitation in ontology engineering.

The only mention of an order of processing texts while forming datasets, we found, was [29], which was the combination of random and chronological selection.

[4] By stability, we term the lowest possible probability of de-saturating a sample after adding a new increment of data items.

This report, however, did not reveal how the chosen selection order influenced data saturation. In our work, we experiment with the orders used in [29] and find these not to be the best choice. In difference to [29], we experiment with additional three possible orderings and recommend the best choice after their cross-evaluation. Our set of orderings accounts not only for the times of publication but also for their impact.

Several publications in Information Science researched how a measurable impact of a publication might influence the views and opinions of a knowledge stakeholder community about their subject domain. The potential impact of a publication has most often been measured using citation or co-citation analysis, with citation counts, scores, or frequencies used as measures. One may observe several interesting refinements, relevant to our work, on the top of this mainstream approach. Pohl and Mottelson [31] introduced the measures of readability, novelty, or name-dropping by defining simple rules based on word occurrences. They also demonstrated the correlation of these measures with citation counts over time. Some publications proposed the refinements of citation analysis to measure potential publication impact in the terms of innovativeness. Schneider and Costas [32] used the Characteristic Scores and Scales (CSS) approach, by Glänzel and Schubert [33], to cluster papers based on citation distribution partitions. Based on this clustering, they filtered out potential highly cited "followers" (false positives) from real candidates to be "breakthrough" papers (true positives). Savov et al. [34] proposed an innovation score based on "how much the predicted publication year is ahead of or behind the actual publication year, which reflects whether the paper covers more topics researched by papers published in the past or more of its topics are covered by future papers". They used a Support Vector Machine (SVM) classifier and latent topic distributions to predict publication timestamps.

While having been aware of the interesting refinements of citation analysis towards accounting for measuring the innovativeness of publication, we noted that these refinements require substantial additional computational overhead. Therefore, we chose to stay with the basic citation frequency measure for document impact in the presented work.

2.8 Automated Term Extraction Methods

The key component technology for our processing pipeline is ATE. Therefore, an informed choice of a method for performing ATE in our experiments was essential. This choice was done based on the review and cross-evaluation of ATE methods and tools by Kosa et al. [35].

Mainstream approaches to ATE [36] processed texts in two consecutive phases: linguistic and statistical. Linguistic pipelines [17] included POS tagging or phrase chunking; filtering out stop words; and restricting candidate terms to n-grams, which could be nouns or noun phrases, adjective-noun and noun-preposition-noun combinations. Statistical pipelines measured the ranks (or scores) of the character strings that were term candidates [36]. These measures [36, 37] focused either on the collocation strength of the units that comprise a term—"unithood" measures; or on the association strength of a term to domain concepts—"termhood" measures.

For "unithood", the following measures were reported as applicable: mutual information [38–40], log likelihood [39–41], *t*-test (*t*-score) [39]; variants of entropy [42]; paradigmatic modifiability and its variants [43]. For measuring "termhood", either the term-frequency-based (unsupervised) or reference-corpora-based (semi-supervised) approaches were used. The most popular frequency-based measures were term frequency (TF) [44]; average or total term frequency (ATF or TTF) [44]; TTF combined with inverse document frequency (TF-IDF) [45, 46]; residual IDF (RIDF) [47, 48]. The measures based on domain corpora were: PU-ATR measure [49]; weirdness [46]; domain pertinence [50]; and domain relevance [51]. Hybrid approaches were also developed to combine "unithood" and "termhood" aspects in one measure. A representative measure for these hybrid approaches was *C*/*NC*-Value [52]. *C*/*NC*-Value-based measures and corresponding ATE methods were further exploited and refined in many works. A few topical examples of these methods could be mentioned. Lossio-Ventura et al. proposed the extension of *C*-Value for single-word terms [53]. Fahmi et al. refined *C*-Value via the use of known terms [43]. GlossEx measure that refined *C*-Value using term confidence information was proposed in [54].

Linguistic pipelines were operationalized very similarly in various ATE methods. One difference was that some of these included the filtering out of stop words and some did not. One of the valid alternatives to cope with the stop words was filtering these out at a cut-off step, after statistical processing. Furthermore, some reported statistical pipelines were split into two consecutive sub-phases of term candidates scoring and ranking. Scoring reflected the likelihood of being terms. Known scoring methods were based on (c.f. [49]) measuring occurrences frequencies (including word associations), assessing occurrences contexts, using reference corpora, e.g. Wikipedia (as in the PU-ATR measure) [49], domain and topic modelling [1, 55, 56].

The cut-off step retained the top-scored candidates, down to a threshold, and thus distinguished significant terms from insignificant (or non-) terms. The scores might have come from one or several scoring algorithms that operationalized different approaches: weighted linear combinations [57]; voting [7, 36]; or (semi-)supervised learning [58]. In our set-up, we perform the cut-off step after computing scores and follow [7, 59] to retain significant terms based on computing a simple majority vote. Hence, ATE solutions performing cut-offs while scoring are irrelevant for us.

The cross-evaluations of the most popular ATE methods and corresponding software solutions were [36, 49]. Based on these reports of performance assessments, ATE methods are compared in Table 2.1. In the table, the comparisons are done using the criteria applicable for the selection of the best appropriate external method for our instrumental software pipeline for measuring terminological saturation.

After analysing the characteristics and performance measurements of different ATE methods, collected in Table 2.1, we support the conclusion of Zhang et al. [36], who stated that *C*-Value is the most reliable method, compared to the rest of the

Table 2.1 The comparison of ATE methods based on the reference implementations and cross-evaluations by Astrakhantsev [49] and Zhang et al. [36]

Method	Domain-independence (±)	Super-vizion (U/SS)	Measure(s)	Term significance	Cu*t*-off (±)	Precision (GENIA; average)	Run Time (related to *C*-value method)	Tool
TTF	+	U	Term (Total) frequency	+	–			ATR4S
						0.70; 0.35	0.34	JATE
ATF	+	U	Average term frequency	+	–	0.71; 0.33	0.37	ATR4S
						0.75; 0.32	0.35	JATE
TTF-IDF	+	U	TTF + Inverse document frequency	+	–			ATR4S
						0.82; 0.51	0.35	JATE
RIDF	+	U	Residual IDF	–		0.71; 0.32	0.53	ATR4S
						0.80; 0.49	0.37	JATE
C-Value	+	*U*	*C-Value, NC-Value*	+	–	*0.73; 0.53*	*1.00*	*ATR4S*
						0.77; 0.56	*1.00*	*JATE*
Weirdness	±	SS	Weirdness	–		0.77; 0.47	0.41	ATR4S
						0.82; 0.48	1.67	JATE
GlossEx	+	SS	Lexical (Term) cohesion, domain specificity	–				ATR4S
						0.70; 0.41	0.42	JATE

(continued)

Table 2.1 (continued)

Method	Domain-independence (±)	Super-vizion (U/SS)	Measure(s)	Term significance	Cu*t*-off (±)	Precision (GENIA; average)	Run Time (related to *C*-value method)	Tool
TermExtractor	+	SS	Domain pertinence, domain consensus, lexical cohesion, structural relevance	–	+			ATR4S
						0.87; 0.46	0.52	JATE
PU-ATR	–	SS	*NC*-value, domain specificity	–	+	0.78; 0.57	809.21	ATR4S
								JATE

Comments to Table 2.1:
Domain Independence: "+"—the method is domain-independent; otherwise—"–", which is either been claimed as domain dependent or been evaluated on one subject domain. We are interested in domain-independent methods.
Supervision: "U"—unsupervised; "SS"—semi-supervised. We are interested in unsupervised methods.
Term Significance: "+"—the method returns a significance score for each retained term; "–"—scores are not returned for terms or the method cuts-off insignificant terms itself. We are interested in receiving significance score and do cut-offs later.
Cut-off: "+"—the method does cut-off and returns only significant terms; "–"—the method does not do cut-offs. Only "-" are relevant for our setting.
Precision and Run Time: The values in the table rows are based on the comparison of the cross-evaluation experiments—Astrakhantsev [49] versus Zhang et al. [36]. Empty cells in the rows mean that the measurements for this method in this experiment were not provided. Astrakhantsev [49] used ATR4S software tool and cross-evaluated 13 different methods on five different datasets, including GENIA dataset [60]. Zhang et al. [36] used JATE 2.0 software and cross-evaluated 9 different methods on two different datasets, including GENIA. Therefore, the results on GENIA could be regarded as a baseline for comparing the precision of ATE. Two values are given for each reference experiment: precision on GENIA and average precision. Both Astrakhantsev [49] and Zhang et al. [36] experimented with the *C*-Value method, which, on average, appeared to be the slowest among the unsupervised methods, as reported in [36]. Therefore, the execution times for *C*-Value are used in Table 2.1 as common denominator for normalizing the rest of the measurements in the **Run Time** column.

methods that were used for cross-evaluation. The *C*-Value method yielded consistently good ATE precision on the two different mixes of datasets [36, 49]. Additionally, it needs to be noted that *C*-Value was one of the slowest in the group of unsupervised and domain-independent methods. However, it did not consume much more computation time than the rest of the evaluated methods. Therefore, its performance was comparable to the faster ones. Though having been not very fast, *C*-Value still outperformed the evaluated domain-specific methods, sometimes significantly (c.f. PU-ATR in Table 2.1). Based on these considerations, we chose *C*-Value as the baseline method for our terminological saturation detection and measurement framework.

2.9 Software Implementations of ATE Methods

For choosing the software tools that implement the *C*-Value method for ATE, the descriptions of term extraction tools at several Web resources were checked, like http://inmyownterms.com/terminology-extraction-tools/ or https://en.wikipedia.org/wiki/Terminology_extraction. In addition to the reference implementations mentioned before, ATR4S and JATE 2.0, the following freely available ATE software tools, as outlined in Table 2.2, were identified.

It was decided not to consider ATR4S and JATE 2.0 because it was not fully clear how to extract the *C*-Value method implementation from these suites. From the freely available software (c.f. Table 2.2), the two relevant tools were shortlisted—NaCTeM TerMine [52] and UPM Term Extractor [62]. These tools implemented the *C*-Value method and were subject domain neutral. For choosing the best appropriate tool from those two for our terminological saturation measurements, their cross-evaluation was performed by Kosa et al. [35], which is reported in Sect. 5.4.

2.10 Text Similarity Measurement

The methods for text or character string similarity, or distance, measurement[5] were (c.f. [63, 64]) classified as belonging to the following groups, distinguished by the ideas for looking at differences:

- Focusing on the differences in characters and character sequences
- Comparing textual tokens
- Finding differences in the terms that are present in the compared texts.
- Looking at the differences or commonalities in the compared text corpora; or
- Analysing differences in the sets of synonyms (synsets), corresponding to the lexical units in the compared texts.

[5] In this context, we do not distinguish a method and a measure. A method is understood as a way to implement the corresponding measure function.

Table 2.2 Freely available software tools for ATE (listed alphabetically)

Name/owner	Link	Short description	Algorithm /metric	Domain	Constraints
BioTex / LIRMM	http://tubo.lirmm.fr/biotex/	Extracts biomedical terms from natural-language texts		Bio-medical	Domain-specific
FiveFilters /Medialab-Prado	http://fivefilters.org/term-extraction/	Extracts terms using a web service; relies on a PHP port of Topia's Term Extraction	Occurrence (TTF) and word count in a term	Neutral	Web service, size of text constrained
TaaS (TaaS EU Project)	https://term.tilde.com/	Identifies term candidates in documents and extracts these automatically	Frequency-based	Neutral	Does not provide term significance scores
TerMine/NaCTeM	http://www.nactem.ac.uk/software/termine/	*Extracts terms from plain English texts and provides the Batch mode (access to be applied to for non-UK academic users)*	*C-Value*	*Neutral*	*The service requests to avoid heavy bulk processing*
TermFinder/Translated.net	https://labs.translated.net/terminology-extraction/	A Web application that extracts terms from the inserted text. Compares the frequency of words in a given document with their frequency in the language (generic corpus)	Poisson statistics, Maximum Likelihood Estimation and IDF	Requires language corpus	Returns the score of a term as a numeric value (%)
TBXTools [61]/Universitat Oberta de Catalunya	https://sourceforge.net/projects/tbxtools/	A Python toolset using NLTK (Natural Language Toolkit)	TTF	Neutral, multilingual, requires language corpus	Deletes n-grams with stop words
UPM Term Extractor [62]/Dr Inventor EU project	https://github.com/ontologylearning-oeg/epnoi-legacy	*A Java software for extracting terms and relations from scientific papers*	*C-Value*	*Neutral*	*Takes text input data of at most 15 Mb*

Some authors (c.f. [64]) extended this methodological typology by either proposing the combinations of several ideas in one measure, as in hybrid approaches or applying fuzzy matching between tokens. A brief summary of the features of the mentioned groups of methods is offered immediately below. A finer-grained overview of the features, for individual methods within the groups, is provided in Table 2.3.

Character (sequence)-based methods compare characters and their sequences in text strings. Sequence information is accounted for by finding differences in the orders of characters. Basic ideas in this group of methods are measuring common character sequences; edit distance; the number and order of the common characters in the compared text strings. The methods in this group are more often used to compare short textual strings.

Token-based methods model a string of text characters as a set of tokens. The tokens could be individual characters, character n-grams with a differing value of n, or separate words as tokens with varying numbers of words n. Similarity or difference are measured by relating the size of the overlap to the length of text. Though the methods in this group are more often used to compare short textual strings, these could be applied to longer texts if the tokens are appropriately chosen.

Term-based methods use the same idea as token-based measures but use different tokens. For this group of methods, appropriate tokens are terms, which are word n-grams with possibly varying n. One more difference is that the weights of the terms are used, that point to their significance, e.g. the frequencies of terms occurrences. The methods in this group are better applicable to longer texts, like documents or textual datasets, where looking at terms might make more sense.

Corpus- or **synset-based** (a.k.a. knowledge-based) **methods** measure the similarity between lexical units (words or word n-grams) using the statistics collected from large textual corpora. Synset-based approaches acquire additional information from lexical (semantic) networks, for example, WordNet [65]. Lexical information is further used to assess similarity between the words in the corpus. Both approaches are overly computationally expensive, though may be applied, for example, to decide what term candidates to cut off. The methods in this group could not be applied for terms grouping based on the similarity of the involved terms, as in our pipeline grouping is optionally performed (c.f. Chap. 4) after significant terms have already been retained. Hence, the methods in this group are marginally relevant for our research purposes.

A more detailed look at the most popular text or textual string similarity measures (SSM) grouped by method types, is offered in Table 2.3. This overview is not complete as we limit our analysis to the methods that might potentially be relevant for our research and avoid the repetitions in underlying ideas or principles.

An expansion-based framework for efficient measuring textual string similarities with an account for the synonyms was presented in [77]. This proposal is relevant for our research as synonyms are the term candidate strings to be considered for terms grouping. As a confirmation of and complement to our selection of the reviewed methods in Table 2.3, Lu et al. [77] mentioned that there was a rich variety of textual string similarity measures available in the literature. These

Table 2.3 An overview of the sample of potentially applicable text similarity (distance) methods

SSM, source	Description	Specifics	Relevance	
			Terms similarity	Terminological difference
Character- or character sequence-based methods				
Longest common substring [66]	Common-character-sequence-based measure	Returns the length of the longest common substring (integer)	Moderate	Irrelevant
Levenshtein distance [67]	Edit-distance-based measure	Returns the number of required edits (integer)	Marginal	Irrelevant
Hamming distance [68]	Edit-distance-based measure	Compared strings have to be of equal length	Marginal	Irrelevant
Monger-Elkan distance [69]	Edit-distance-based measure	Returns the number of required edits (integer)	Marginal	Irrelevant
Jaro distance [70]	Counts the minimal number of single character transforms in one string needed to arrive at the other string	Returns a normalized real value from [0, 1]	Good	Irrelevant
Jaro-Winkler distance [71]	Refines Jaro measure by using a prefix scale to prioritize the stings that match at the beginnings	Returns a normalized real value from [0, 1]	Good	Irrelevant
Token-based methods				
Sørensen-Dice coefficient [72, 73]	Counts the ratio of identical character *bi*-grams to the overall number of *bi*-grams in both strings	Returns a normalized real value from [0, 1]	Good	Irrelevant
Jaccard similarity [74]	Counts the ratio of the cardinalities of the intersection and union of the character *uni*-grams	Returns a normalized real value from [0, 1]	Good	Irrelevant

(continued)

Table 2.3 (continued)

SSM, source	Description	Specifics	Relevance	
			Terms similarity	Terminological difference
Cosine similarity [64]	Relates the size of overlap in character *uni*-grams to the square root of the sum of the squared total numbers of *uni*-grams in both strings	Returns a normalized real value from [0, 1]	Marginal (computationally hard)	Irrelevant
Term-based measures				
Euclidian distance [75]	Measures traditional Euclidian distance in an *n*-dimensional metric space (of non-negative reals)	Applicable to documents and textual datasets; returns a real value	Irrelevant	Relevant
Cosine similarity [76]	Computes a cosine between two vectors in the term space; vectors are specified by term weights (e.g. TF of *C*-Value)	Applicable to documents and textual datasets; returns a normalized real value from [0, 1]	Irrelevant	Marginal
Pearson correlation [75]	Computes Pearson correlation for a pair of vectors in the term vector space	Applicable to documents and textual datasets; returns a normalized real value from [−1, 1]; it equals 1 when the vectors are fully identical	Irrelevant	Marginal
Manhattan (block) distance [63]	Measures the distance to be travelled to get from one data point to the other if a grid-like path is followed	Applicable to documents and textual datasets; resembles the thd measure (Chap. 3)	Irrelevant	Relevant

measures comprised character *n*-gram similarity [78], Levenshtein distance [67], Jaro-Winkler measure [71], Jaccard similarity [74], TF/IDF-based cosine similarity [79], and Hidden-Markov-Model-based measure [80].

2.11 Efficient Strings Matching for Searching Nested Terms

It is important to possess an efficient way to match multiple textual strings. Such matching is the operation that is heavily exploited in the statistical pipelines for ATE. This problem is complex and it was addressed by many authors. In one of the recent overviews by Chowdhury and Farrell [81] several textual string matching algorithms were reviewed with respect to the use for exploring the occurrences (nestings) of term strings in the longer term strings. Based on this overview and analysis, Chowdhury and Farrell proposed their `TBI` algorithm. The summary of the comparative overview [81] is given in Table 2.4. In addition to the output of [81], the performance assessment of our baseline implementation (Sect. 4.5) was added in the last row.

Chowdhury and Farrell [81] admitted that their review was not complete. They mentioned some other relevant textual strings matching algorithms, for example, Karp and Rabin [82], but pointed out that the performance of these additionally mentioned algorithms was similar or worse compared to the overviewed. Therefore, as per [81], the methods listed in Table 2.4 form a representative sample.

The external ATE software integrated into our baseline processing pipeline, UPM Term Extractor [62], exploited a variant of [36] for nested terms search. It could be seen, however, that according to the theoretical estimations of the computational complexity (Table 2.4), a key-value store based approach [36] was not an optimal choice. The algorithm proposed by Aho and Corasick [86] was assessed as more computationally efficient. Therefore, it was decided to use the Aho-Corasick algorithm in *C*-Value computation (c.f. Sect. 4.3.1) for our optimized processing pipeline.

2.12 Research Gaps and Motivation

In Sects. 2.2, 2.3, 2.4, 2.5, 2.6, 2.7, 2.10 and 2.11 several research gaps (**RG**) were identified, that motivated our work as follows:

- **RG1: Potential bias in the interpretations of a subject domain.** In mainstream ontology engineering, there is a lack of an objective measure for the representativeness of the expert group for requirements elicitation and, therefore, for the completeness of the requirements elicited from the available experts.

This research gap motivated us to: look into ontology learning for bypassing human experts at the requirements elicitation phase; substitute expert interviews

Table 2.4 Performance comparison of the algorithms for nested terms search, after [81]. Asymptotic notation $O(.)$ was used for comparison

Implementation source	Data Structure, source	Computation complexity			Parameters
		Indexing	Search	Total	
[36]	key-value store, a.k.a hash table [83]	$O(v^2)$	$O(m^2)$	$O(v^2 + v * m^2)$	v—number of term candidate strings s—average number of nesting term strings m—number of tokens in the input (nested) term, $v \gg m$
[84]	Trie [85], where words are nodes	$O(v * t)$	$O(m^2)$	$O(v * t + v * m^2)$	v—number of term candidate strings t—average number of words for a term, $v \gg t$
[86]	Trie [85], where words are nodes	$O(v * t)$	$O(m + L)$	$O(v * t + m + L)$	v—number of term candidate strings t—average number of words for a term, $v \gg t$ m—number of tokens in the input (nested) term, $v \gg m$ L—total length of all terms
TBI, [81]	Term store composed of two hash tables [81]	$O(v * \log(v))$	$O(m^2)$	$O(v * \log(v) + m^2)$	v—number of term candidate strings m—number of tokens in the input (nested) term, $v \gg m$

(continued)

Table 2.4 (continued)

Implementation source	Data Structure, source	Computation complexity			Parameters
		Indexing	Search	Total	
Exhaustive search, [7]	An array of term strings	0	$O(v^2 * m)$	$O(v^2 * m)$	v—number of term candidate strings m—number of tokens in the input (nested) term, $v \gg m$

by extracting terms from relevant text collections; explore qualitative research methodologies for insights on how to ensure that the sample of texts is representative.

- **RG2: Lack of a measure for the representativeness of a document collection**. In ontology learning from texts, there is a lack of measures of the representativeness of a document collection for sufficiently covering the interpretations of a subject domain by the knowledge stakeholders. Hence, there is no guarantee that the used text collection is representative, even if it is very big in volume. Even more, it might happen that only a small part of a very big text collection is sufficiently complete to cover domain knowledge satisfactorily.

This research gap motivated us to develop such a measure (thd) and prove its metric properties. In our approach, this metric is used for the detection of terminological saturation and building terminological core sub-collections, as presented in Chap. 3. The rationale for focusing on the phenomenon of terminological saturation is that having a method to circumscribe a terminological core sub-collection within the entire collection would help substantially decrease the effort needed for ontology learning from these texts.

- **RG3: Lack of the scalability of ATE for processing full-text collections of industrial volumes**. In research on learning ontologies based on scholarly data, scalable approaches, like Klink-2, process only the metadata of documents. A scalable approach that processes full texts from industrial size collections containing dozens of thousands to millions of papers is not available, to the best of our knowledge. Scaling ATE is not trivial as the known methods are computationally expensive in the terms of their run time versus the volume of input text. The *C*-Value method, that we have chosen for our terminological saturation measurement pipeline (Sects. 5.3 and 5.4), is more expensive than the other unsupervised and domain-neutral methods. Furthermore, ATE implementations are often constrained[6] in the volume of input text.

To narrow this research gap, we were motivated to decrease the volume of processed data, while keeping it representative and to increase the performance of the *C*-Value method for ATE. The decrease of the volume was sought via developing the method of extracting terminological core sub-collections (Sect. 3.1); configuring the input data for this method in a way to make the outputs more compact and stable (Sects. 5.2 and 5.5). For improving the performance of the *C*-Value method, we sought an approach that correctly partitions incrementally enlarged document sub-collections to allow parallel processing of small data chunks (Sect. 3.6); and uses fast textual strings matching algorithm in the statistical part of the ATE pipeline (Sect. 4.4.1).

[6] For example, the UPM Term Extractor, used as a baseline ATE module in our experiments, implements the *C*-Value method and constrains the volume of input text by 15 Mb in the used hardware configuration.

- **RG4: Lack of a method and criterion for detecting terminological saturation**. In qualitative research, though exploiting data saturation, there is no objective formal method and criterion for finding out if saturation exists and detecting it.

The lack of a method motivated us to propose the use of a successive approximation process and respective saturation criterion (Sect. 3.1) and prove the existence theorem in Sect. 3.5.

- **RG5: Lack of knowledge about the most appropriate ATE method and software for detecting terminological saturation**. Many methods are known in ATE that are compared based on their extraction quality (precision, recall, F-measure). However, there is no information about those methods that helps judge their quality in the context of detecting terminological saturation.

Due to the availability of numerous ATE methods and their implementations, we did not have the motive to invent and implement yet one more to those. Instead, we were motivated to experimentally cross-evaluate the relevant candidates in our terminology saturation detection pipeline and select the best appropriate, as reported in Sects. 5.3 and 5.4.

- **RG6: Lack of knowledge of an optimal ordering of documents for processing**. In qualitative research, there is also no rule for ordering interview documents that increases the effectiveness and efficiency of their processing.

The lack of a rule for ordering documents motivated us to look at and cross-evaluate different possible document orderings based on time stamps and measured document impacts. This part of work is reported in Sect. 5.5.

- **RG7: Lack of knowledge whether term grouping yields more compact and stable saturated term-sets**. Known ATE methods do not group together the terms that are fully or partially similar to each other. Partial similarity is exploited, in some approaches (c.f. [87]), for taxonomy generation which happens later in the ontology learning pipeline (c.f. [16]).

This research gap motivated us to look at and cross-evaluate several short string similarity measures, define terms similarity thresholds that help improve terminological saturation when fully and/or partially similar terms are grouped. The results are presented in Sects. 4.6 and 5.6.

2.13 Research Questions and Objectives

In this Section, based on the analysis of the related work and identified research gaps (**RG1—RG7**, Sect. 2.12), we aim at formulating our research questions and objectives. We approach this aim by presenting, in Sect. 2.13.1, our vision of the approach that might narrow the discovered research gaps. In this concise but very shallow presentation, we put upfront several research questions in different contexts within the proposed approach. These questions and anticipated ways to look for the answers are elaborated in Sect. 2.13.2. Finally, the objectives of our research are formulated in Sect. 2.13.2.

2.13.1 Envisioned Approach for Terminological Saturation Detection and Measurement

We seek the solution for the automated extraction of representative term-sets using an iterative successive approximation approach. An iteration of this envisioned method is pictured in Fig. 2.1. The idea is borrowed from the Grounded Theory method by Glaser and Strauss [4]. We apply this idea to building, iteratively, a terminological core sub-collection of documents, which contains a saturated term-set, representative for the corresponding subject domain.

We further refine the heuristic by Glaser and Strauss [4] by proposing the bits that are missing in the Grounded Theory method and proving their validity.

These bits are:

- The formal criterion of reaching terminological saturation, including a measure, that allows finishing the process
- The conditions under which terminological saturation exists.

We propose that the process of successive approximation starts with an empty set of documents—the dataset D_0. At every iteration of the method, for example, the i-th, several (*inc*) new documents are taken from the input collection (*DC*), pre-processed to a plain text file, and appended to the dataset (D_{i-1}) that has been processed at the previous iteration ($i-1$). This forms the dataset (D_i) for the i-th iteration. Due to this procedure, the datasets incrementally grow in the iterations. We suppose that, while growing, the datasets successively become closer to the dataset (D_{sat}) corresponding to the terminological core sub-collection of documents (*TCSC*). The saturated term-set (T_{sat}) is finally acquired from D_{sat}.

To evaluate how close we are to D_{sat}, we rely on measuring terminological difference between the sets of significant terms retained from the term candidates extracted from the datasets in successive iterations. In the i-th iteration, terminological difference (thd_i) is measured between the sets of retained significant terms T_i and T_{i-1}.

For generating T_i in the i-th iteration, we propose the following steps:

- Extract the set of term candidates, with their significance values, from D_i
- Compute the cut-off threshold eps_i to separate significant terms from insignificant terms
- Retain the term candidates having significance values above eps_i in T_i as significant terms
- For detecting terminological saturation, we propose to observe thd_i versus eps_i and stop the process when thd_i reliably goes below eps_i.

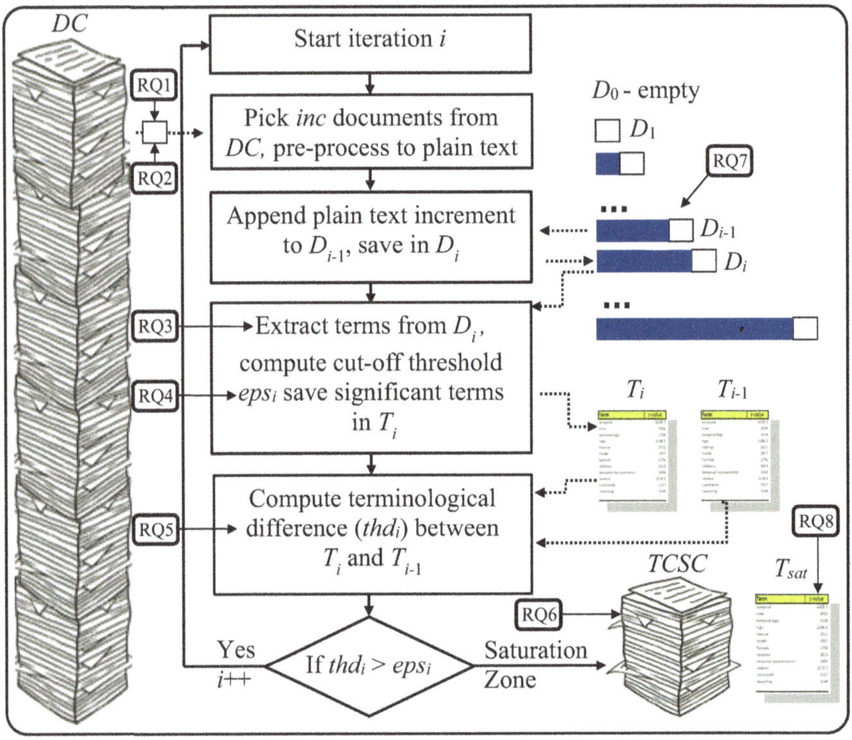

Legend:

	DC	– a domain-bounded document collection
	inc	– the number of documents, picked from the DC, that are appended to a dataset at any iteration
	D_i	– the plain text dataset for the *i*-th iteration
	T_i	– the set of significant terms retained from D_i
	eps_i	– the cut-off threshold that separates significant from insignificant terms
	thd_i	– the measure of terminological difference between the two arbitrary sets of terms
	TCSC	– the terminological core sub-collection of documents
	T_{sat}	– the saturated set of terms retained from the dataset corresponding to *TCSC*
	⟶	– control flow
	·······▸	– data flow
	RQ...	– a pointer to the context of a research question (Section 2.13.2)

Fig. 2.1 The vision of a method for extracting saturated term-sets from a domain-bounded document collection

2.13.2 *Research Questions and Objectives*

As pointed out in Fig. 2.1, the answers to several research questions, **RQ1** to **RQ8**, are required for elaborating the outlined approach into a method. In this Section, the anticipated ways to approach the answers are outlined in the contexts of individual questions. The section is summarized by Table 2.5 that offers the pointers to the chapters, sections, and artefacts within the book.

RQ1 (document ordering): *Which documents in the unprocessed part of the collection are to be picked to form the next increment for processing?*

Documents in a collection differ from each other by their terminological "footprint" on the domain. The differences are caused by:

- The changes of the terminological focus of the knowledge stakeholder community over time—so to say, their terminological drift in time
- The impact of the documents on the terminology shared among the knowledge stakeholder community in the domain.

In this work, the answer to this question is given by experimentally cross-evaluating five possible document orderings. These orderings are based on the differences in the time of documents publications and their anticipated impacts on terminology in the domain. The review of the related work on this point is given in Sect. 2.7. The result of the cross-evaluation and our recommendation on the most balanced ordering are reported in Sect. 5.5.

RQ2 (increment size): *Does the size of the document increment (inc) matter? Is there any optimal number of documents for an increment that makes terminological saturation achievable and stable in the least number of iterations?*

Datasets might be generated using smaller or larger increments. Straightforwardly, the smaller is number of documents in an increment, the higher is the number of required iterations. On the other hand, the smaller is the increment, the higher might be the precision of the extraction of a terminological core sub-collection. Hence, it might be reasonable to find an optimal size of an increment for balancing performance and precision. In our work, the answer to this question is given formally by proving the Corollary 3.6 to Theorem 3.5 on the equality of cv and $mpcv$ in Sect. 3.7.

RQ3 (ATE method): *Which ATE method is the most suitable for the proposed successive approximation approach?*

In the related work, the methods for ATE and their software implementations were compared by the quality of term extraction on different datasets (c.f. Sect. 2.8). This criterion, though important, is not fully relevant for our context as terminological difference is measured between the term-sets extracted by the same method—therefore characterized by the identical extraction quality. Hence, additional aspects, to quality, need to be taken into account for choosing the best appropriate ATE method implementation in our work. The answer to this question is therefore given via the cross-evaluation of a few software implementations, shortlisted due to their quality on the mix of different datasets. Cross-evaluation is done experimentally (Sects. 5.3 and 5.4) by using the shortlisted software in the processing pipeline outlined in Fig. 2.1, and comparing the results.

RQ4 (cut-off threshold): *How would the cut-off threshold for significant terms (eps) be chosen? What is the rationale for this choice?*

The rationale for cutting off the insignificant part of a term-set has been proposed by Tatarintseva et al. [7], who based it on the public election voting allusion. Indeed, a winner in a public election is the one who gets the simple majority of votes of the

Table 2.5 Research objectives, questions, and answers

Research objective	Research question	Answer	
		Sections	Artefact
RO1 (formal framework)	**RQ2** (optimal increment size)	3.6	**Corollary** 3.6. Size of a partial collection
	RQ4 (cut-off threshold)	3.1	**Definition** 3.4. Individual term significance threshold (*eps*)
	RQ5 (terminological difference computation)	3.3 and 3.4	Terminological difference metric function (*thd*)
	RQ6 (existence of terminological saturation)	3.5	**Theorem** 3.3. Sufficient conditions for the existence of terminological saturation
	RQ7 (optimization and scalability of ATE)	3.6	**Theorem** 3.5. Equality of cv and $mpcv$
RO2 (algorithmic suite)	**RQ1** (document ordering)	4.3.2	Configuration and dataset generation (**GDS**)
	RQ4 (cut-off threshold)	4.5 and 4.6.4	Baseline algorithm (**THD**) Refined algorithm (**R-THD**)
	RQ5 (terminological difference computation)	4.5 and 4.6.4	Baseline algorithm (**THD**) Refined algorithm (**R-THD**)
	RQ7 (optimization and scalability of ATE)	4.4	Efficient algorithm for computing (partial) *C*-Values (**AC-CV**) Algorithm for merging partial *C*-Values (**MPCV**)
	RQ8 (term grouping)	4.6	Selection of terms similarity measures Terms similarity thresholds Algorithms for term grouping (**STG**, **M-JR**, **M-JW**, **M-JA**, **M-SD**) Refined terminological difference measurement algorithm (**R-THD**)

(continued)

Table 2.5 (continued)

Research objective	Research question	Answer	
		Sections	Artefact
RO3 (experimental evaluation and validation)	**RQ3** (ATE method and software)	5.3 and 5.4	*C*-Value method implemented in the UPM Term Extractor software
	RQ9 (correctness of the pipeline)	5.3	The developed pipeline (method, algorithms, and software) computes correct results in the boundary cases
	RQ1 (document ordering)	5.5	Descending citation frequency ordering (**dcf**)
	RQ7 (optimization and scalability of ATE)	5.7	Experimental proof of the *h*1 hypothesis of Theorem 3.5. Experimental validation of the efficacy, domain neutrality, and scalability of the optimized pipeline
	RQ8 (term grouping)	5.6	Experimental evaluation of terms grouping
RO4 (practical use)	**RQ10** (use cases)	6.1 and 6.2	The use of the developed technology in an industrial project (Sect. 6.1). The use of the developed technology in an academic curriculum (Sect. 6.2)
	RQ11 (practical implications)	6.3 and 6.4	Practical benefits of the use of the developed technology for industrial adopters (Sect. 6.3). Potential business scenarios in scholarly publishing (Sect. 6.4)

electorate, which is, minimally, 50 percent, plus one vote. In this work, we follow this approach and elaborate the formalism for it in Chap. 3.

RQ5 (terminological difference): *How to compute terminological difference (thd) between two bags of terms, in which terms are supplied with their significance score values? What are the properties of this function?*

Evidently, there are two technical parts in this question.

The first is how to compare terms, which are short textual strings? Several measures for string similarity are known from the related work (c.f. Sect. 2.10). In addition, several algorithms for fast textual strings matching are available (c.f. Sect. 2.11). In this work, we explore several measures to match term strings fully or partially in two contexts: term grouping (Sects. 3.7, 4.4, and 5.6); and *thd* computation (Sect. 4.5).

There is no formal answer in the related work, however, about how to measure the difference between the sets of retained significant terms, given that every term is supplied with a real number—its significance score. Tatarintseva et al. [7] proposed an algorithm for this measurement. Based on their work, we formally define the *thd* measure in Sect. 3.3 and prove its metric properties in Sect. 3.4.

RQ6 (existence of terminological saturation): *How to check if terminological saturation is achievable, given a particular document collection? What if thd went below eps in some iteration i and goes up, beyond eps, in iteration j > i? How to make sure that saturation is stable after its detection in iteration i?*

The answer to this group of questions is sought formally by proving the existence Theorem 3.3 in Sect. 3.6.

RQ7 (optimization and scalability of ATE): *Could the ATE method, used in the proposed solution for terminological saturation detection, be optimized to scale well to the volumes of realistic document collections of industrial size?*

ATE is known to be expensive with respect to computation time (c.f. Sect. 2.8). Furthermore, its memory requirements grow quickly with the volume of text to be processed. Hence, ATE software implementations are slow and constrained in the volume of text taken in. This feature of the available ATE software is especially restrictive if the datasets are incrementally enlarged and processed iteratively, as it is in the proposed method for saturated terminology extraction. A useful idea to overcome this deficiency in scalability might be reverting the order of operations in the pipeline, as pictured in Fig. 2.2.

Indeed, if the used ATE method allows adequately merging the bags of retained significant terms then it might be applied only to the increments but not to the full incrementally growing datasets, as pictured in Fig. 2.2b. As the partition datasets of a collection comprising only the increments are small, both constraints of ATE might be waved. Using this idea, the answers to this group of questions are approached by:

- Proving that the *C*-Value method, that is chosen for our processing pipeline, allows term extraction with further merging in Sect. 3.6
- Developing the relevant algorithms in Sect. 4.4

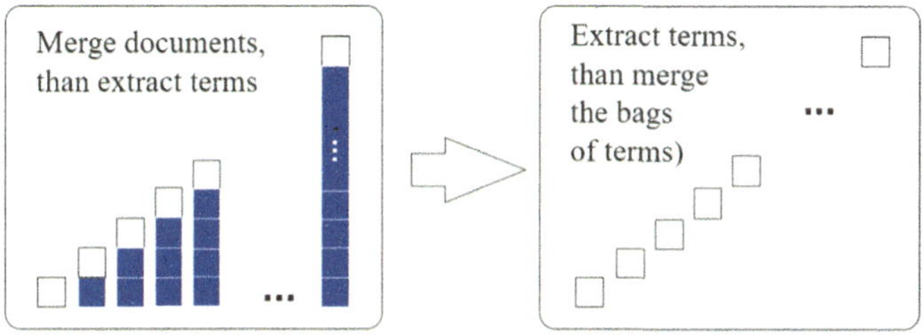

(a) Incremental approach (Sect. 2.13.1)

(b) Optimized approach (Sect. 3.7 and 4.3)

Fig. 2.2 Reverting dataset incremental enlargement and term extraction for the better scalability and increased performance of ATE

- Evaluating, experimentally, the validity, performance, domain neutrality, and scalability of the optimized processing pipeline for saturated terminology extraction in Sect. 5.7.

RQ8 (term grouping): *Would the use of similarity thresholds for grouping partially similar terms, while measuring terminological difference, improve terminological saturation by lowering the required number of iterations and making saturation more stable?*

Our deliberation behind this question is that an exact string match gives a more narrow and incomplete result than a semantic match in a pair of terms. Indeed, semantically the same terms may however be represented by slightly different strings in a text. Hence, a way to assess semantic similarity might be more favourable in computing terminological differences than relying on exact string matches.

In our work (Sect. 4.6.2), an approach to rationally choose the thresholds for regarding string similarity as semantic (or term) similarity is sought. The thresholds are chosen for four selected string similarity measures (Sect. 4.6.1). Further, an algorithm for similar terms grouping is developed, that uses the chosen string similarity measures and terms similarity thresholds (Sect. 4.6.3). Based on its use, the baseline **THD** algorithm [7] for measuring terminological difference between two bags of retained significant terms (Sect. 4.6.4) is refined. Finally, the refined **THD** algorithm and the algorithms for string similarity measurement are experimentally evaluated in Sect. 5.6.

RQ9 (correctness of the developed pipeline): *Do the developed method, algorithms and software for terminological saturation measurement produce correct results in the boundary cases of document collections: (i) the one in which steady saturation should be detected very quickly; and (ii) the one in which saturation cannot be detected?*

To answer this question experimentally, the two synthetic document collections for the outlined boundary cases are assembled: 1DOC and RAW (Sect. 5.2.2). The pipeline, developed in Chaps. 3 and 4, is applied to these collections and the results are experimentally checked for correctness in Sect. 5.3.

In addition to the scientific and technical questions, formulated and discussed above, our aim in this research work was to seek the practical use of the proposed method of saturated terminology extraction. In this context, additional two research questions are formulated as follows.

RQ10 (use cases): *Given the method is developed and proven valid, what are the relevant industrial and academic use cases for it?*

To answer this question, we report on the two use case studies: one in an industrial and the other in an academic project in Sects. 6.1 and 6.2 respectively.

RQ11 (practical implications): *What are the potential practical benefits and limitations with respect to the use of the method as a part of information technology? What are the anticipated industrial scenarios for technology transfer?*

The practical benefits for potential technology adopters and the limitations of the developed solution are summarized in Sect. 6.3. Potential business scenarios for applying our technology in scholarly publishing and digital libraries are outlined in Sect. 6.4.

The research objectives in our study are formulated in a way to group the estimated results of the answers to our research questions methodologically and structurally. The objectives and are as follows.

RO1 (formal framework): Our first objective is to develop a theoretical framework covering representations, formalisms, and proofs for all the elements required to answer our research questions. This framework is reported in Chap. 3.

RO2 (algorithmic suite): Given that the formal framework is elaborated, our second objective is to develop the suite of algorithms that operationalize the proposed method for saturated terminology extraction. These algorithms should be further implemented in the software suite. The software should be made ready and available for academic and industrial use. These results are presented in Chap. 4.

RO3 (experimental evaluation and validation): Given that the software suite is developed, our third objective is to evaluate and validate the method for saturated terminology extraction on synthetic and real document collections of scholarly publications belonging to different subject domains. This contribution is reported in Chap. 5.

RO4 (practical use): Finally, our fourth objective is to highlight the practical benefits and outline the limitations of the use of the developed method within the initial phase of ontology learning from texts. In this context, the focus is put on selected industrial and academic application scenarios for technology transfer. Our experience, to date, in the deployments of the developed method and implemented software pipeline in industry and academy is reported. Based on these demonstrator cases, our understanding of the practical benefits for potential technology adopters is summarized. Several potential business scenarios in the scholarly publishing industry, in which the use of the developed technology might be beneficial, are discussed. These aspects comprise Chap. 6.

2.14 Summary

In this Chapter, we reviewed the State-of-the-Art related to the contributions of our work. A systematic review of the related publications has been undertaken for collecting different bits of relevant knowledge in various fields. The analysis of these bits of knowledge helped us better understand the research gaps in our field of study (Sect. 2.12).

With an aim to narrow these research gaps, a vision of the approach for terminology saturation measurement and detection was proposed in Sect. 2.13.1. This vision allowed us to formulate the research questions that needed to be answered in order to transform this visionary approach into the method, further implement it in a software pipeline, systematically evaluate the method, and make it ready for technology transfer. For that, the research questions **RQ1** to **RQ11** have been grouped in the research objectives **RO1** to **RO4** (Sect. 2.13.2).

The foci of the research objectives on the groups of research questions are outlined in Table 2.5. The table also provides the key to locating the answers to the research questions in the reminder of the book.

References

1. Dobrovolskyi H, Keberle N (2018) Collecting seminal scientific abstracts with topic modelling, snowball sampling and citation analysis. In: ICTERI 2018: 14th International conference on ICT in education, research, and industrial applications. Volume I: Main conference, Kyiv, Ukraine, 2018, CEUR-WS, vol 2105, p 179–192
2. Ferrari A, dell'Orletta F, Spagnolo GO, Gnesi S (2014) Measuring and improving the completeness of natural language requirements. In: Salinesi C, van de Weerd I (eds) REFSQ 2014: Requirements engineering: foundation for software quality. Lecture notes in computer science, vol 8396, Springer, Cham, p 23–38. https://doi.org/10.1007/978-3-319-05843-6_3
3. Aldiabat KM, Le Navenec C-L (2018) Data saturation: the mysterious step in grounded theory method. Qual. Rep. 23(1):245–261
4. Glaser BG, Strauss A (1967) The discovery of grounded theory: strategies for qualitative research. Aldine, Chicago IL
5. Guarino N, Oberle D, Staab S (2009) What is an ontology? In: Staab S, Studer R (eds) Handbook on ontologies. International handbooks on information systems. Springer, Berlin, Heidelberg, p 1–17. https://doi.org/10.1007/978-3-540-92673-3_0
6. Studer R, Benjamins R, Fensel D (1998) Knowledge engineering: principles and methods. Data Knowl Eng 25(1–2):161–198. https://doi.org/10.1016/S0169-023X(97)00056-6
7. Tatarintseva O, Ermolayev V, Keller B, Matzke W-E (2013) Quantifying ontology fitness in OntoElect using saturation- and vote-based metrics. In: Ermolayev V et al (eds) Revised Selected Papers of ICTERI 2013. Communications in computer and information science, vol 412. Springer, Heidelberg, p 136–162. https://doi.org/10.1007/978-3-319-03998-5_8
8. Sure Y, Staab S, Studer R (2003) On-To-Knowledge methodology. In: Staab S, Studer R (eds) Handbook on ontologies. International handbooks on information systems. Springer, Berlin, Heidelberg, p 117–132. https://doi.org/10.1007/978-3-540-24750-0_6
9. Gómez-Pérez A, Fernández-López M, Corcho O (2004) Ontological engineering. Springer, London.https://doi.org/10.1007/b97353

10. Pinto HS, Tempich C, Staab S, Sure Y (2004) DILIGENT: Towards a fine-grained methodology for distributed, loosely-controlled and evolving engineering of ontologies. In: de Mántaras RL, Saitta L (eds) ECAI'04: Proceedings of the 16th European conference on artificial intelligence, Valencia, 2004. IOS Press, p 393–397. https://doi.org/10.5555/3000001.3000084
11. Schreiber G, Akkermans H, Anjewierden A, de Hoog R, Shadbolt NR, Van de Velde W, Wielinga BJ (1999) Knowledge engineering and management: the CommonKADS methodology. The MIT Press, Cambridge, MA
12. Suárez-Figueroa MC, Gómez-Pérez A, Motta E, Gangemi A (2012) Ontology engineering in a networked world. Springer, Heidelberg.https://doi.org/10.1007/978-3-642-24794-1
13. Sarasua C, Simperl E, Noy N, Bernstein A, Leimeister JM (2015) Crowdsourcing and the Semantic Web: a research manifesto. Human Comput 2(1):3–17. https://doi.org/10.15346/hc.v2i1.2
14. Maedche A, Staab S (2001) Ontology learning for the Semantic Web. IEEE Intell Syst 16(2):72–79. https://doi.org/10.1109/5254.920602
15. Buitelaar P, Cimiano P, Magnini B (2005) Ontology learning from text: methods, evaluation, and applications. In: Frontiers in Artificial Intelligence and Applications, vol 123. IOS Press
16. Wong W, Liu W, Bennamoun M (2012) Ontology learning from text: a look back and into the future. ACM Comput. Surv. 44(4):20:1–20:36. https://doi.org/10.1145/2333112.2333115
17. Maynard D, Bontcheva K, Augenstein I (2017) Natural language processing for the semantic web. Morgan Claypool. https://doi.org/10.2200/S00741ED1V01Y201611WBE015
18. Kosa V, Chugunenko A, Yuschenko E, Badenes-Olmedo C, Ermolayev V, Birukou A (2017) Semantic saturation in retrospective text document collections. In: Mallet F, Zholtkevych G (eds) ICTERI 2017: Proceedings of Ph.D. symposium. Kyiv, 2017, CEUR-WS, vol 1851, p 1–8
19. Corpas Pastor G, Seghiri Domínguez M (2010) Size matters: a quantitative approach to corpus representativeness. In: Rabadán R, Fernández López M, Guzmán González T (eds) Lengua, traducción, recepción en honor de Julio César Santoyo, León, 2010, Universidad de León, Secretariado Área de Publicaciones y Medios Audiovisuales, p 111–145
20. Salatino AA, Thanapalasingam T, Mannocci A, Osborne F, Motta E (2018) The computer science ontology: A large-scale taxonomy of research areas. In: Vrandečić D et al (eds) ISWC 2018: The Semantic Web, 1996. Lecture notes in computer science, vol 11137. Springer, Cham, p 187–205. https://doi.org/10.1007/978-3-030-00668-6_12
21. Osborne F, Motta E (2015) Klink-2: integrating multiple Web sources to generate semantic topic networks. In: Arenas M et al (eds) ISWC 2015: the Semantic Web, 2015. Lecture notes in computer science, vol 9366. Springer, Cham, p 408–424. https://doi.org/10.1007/978-3-319-25007-6_24
22. Waddington H, White H, Snilstveit B, Hombrados JG, Vojtkova M, Davies P et al (2012) How to do a good systematic review of effects in international development: a tool kit. J Develop Effectiveness 4(3):359–387. https://doi.org/10.1080/19439342.2012.711765
23. Lecy JD, Beatty KE (2012) Representative literature reviews using constrained snowball sampling and citation network analysis. SSRN. https://doi.org/10.2139/ssrn.1992601
24. Ahad A, Fayaz M, Shah AS (2016) Navigation through citation network based on content similarity using cosine similarity algorithm. Int J Database Theor Appl 9(5):9–20. https://doi.org/10.14257/ijdta.2016.9.5.02
25. Medlar A, Ilves K, Wang P, Buntine W, Głowacka D (2016) PULP: A system for exploratory search of scientific literature. In: SIGIR'16: Proceedings of the 39th international ACM SIGIR conference on research and development in information retrieval. ACM, New York, NY, p 1133–1136. https://doi.org/10.1145/2911451.2911455
26. Varela AR, Pratt M, Harris J, Lecy J, Salvo D, Brownson RC, Hallal PC (2018) Mapping the historical development of physical activity and health research: a structured literature review and citation network analysis. Prev Med 111:466–472. https://doi.org/10.1016/j.ypmed.2017.10.020
27. Riazanov A, Voronkov A (2001) Adaptive saturation-based reasoning. In: Bjørner D, Broy M, Zamulin AV (eds) PSI 2001: Perspectives of system informatics. Lecture notes in computer

science, vol 2244. Springer, Berlin, Heidelberg, p 55–61. https://doi.org/10.1007/3-540-45575-2_11

28. Chernyak L, Berenstein A (2006) Method and apparatus for informational processing based on creation of term-proximity graphs and their embeddings into informational units. US patent application publication, No US 2006/0031219 A1, Feb 9
29. Doerre J, Gerstl P, Goeser S, Mueller A, Seiffert R (2002) Taxonomy generation for document collections. US patent, No US 6 446 061 B1, Sep 3
30. Han H, Xu S, Zhu L (2014) Mining technical topic networks from Chinese patents. In: IPAMIN 2014: Proceedings of the 1st international workshop on patent mining and its applications. Hildesheim, 2014. CEUR-WS, vol 1292
31. Pohl H, Mottelson A (2019) How we guide, write, and cite at Chi. In: CHI EA'19: Extended abstracts of the 2019 CHI conference on human factors in computing systems, Glasgow, Scotland. ACM New York, NY, paper no. alt01, p 1–11. https://doi.org/10.1145/3290607.3310429
32. Schneider JW, Costas R (2017) Identifying potential 'breakthrough' publications using refined citation analyses: three related explorative approaches. J Am Soc Inf Sci 68(3):709–723. https://doi.org/10.1002/asi.23695
33. Glänzel W, Schubert A (1988) Characteristic scores and scales in assessing citation impact. J Inf Sci 14(2):123–127. https://doi.org/10.1177/016555158801400208
34. Savov P, Jatowt A, Nielek R (2020) Identifying breakthrough scientific papers. Inf Process Manage 57(2):102–168. https://doi.org/10.1016/j.ipm.2019.102168
35. Kosa V, Chaves-Fraga D, Naumenko D, Yuschenko E, Badenes-Olmedo C, Ermolayev V, Birukou A (2018) Cross-evaluation of automated term extraction tools by measuring terminological saturation. In: Bassiliades N et al (eds) ICTERI 2017: Revised selected papers. Communications in computer and information science, vol 826. Springer, Cham, p 135–163. https://doi.org/10.1007/978-3-319-76168-8_7
36. Zhang Z, Iria J, Brewster C, Ciravegna F (2008) A comparative evaluation of term recognition algorithms. In: 6th international conference on language resources and evaluation, p 2108–2113
37. Korkontzelos I, Klapaftis IP, Manandhar S (2008) Reviewing and evaluating automatic term recognition techniques. In: Nordström B, Ranta A (eds) GoTAL 2008: Advances in natural language processing, 2008. Lecture notes in computer science, vol 5221. Springer, Berlin, Heidelberg. p 248–259. https://doi.org/10.1007/978-3-540-85287-2_24
38. Church KW, Hanks P (1990) Word association norms, mutual information, and lexicography. Comput Linguist 16(1):22–29
39. Wermter J, Hahn U (2005) Finding new terminology in very large corpora. In: Clark P, Schreiber G (eds) 3rd International conference on knowledge capture. ACM, p 137–144. https://doi.org/10.1145/1088622.1088648
40. Daille B (1996) Study and implementation of combined techniques for automatic extraction of terminology. In: Klavans J, Resnik P (eds) The balancing act: combining symbolic and statistical approaches to language. The MIT Press, Cambridge, Massachusetts, pp 49–66
41. Dunning T (1993) Accurate methods for the statistics of surprise and coincidence. Comput Linguist 19(1):61–74
42. Manning CD, Schutze H (1999) Foundations of statistical natural language processing. Bradford Book & MIT Press, Cambridge, MA, London
43. Fahmi I, Bouma G, van der Plas L (2007) Improving statistical method using known terms for automatic term extraction. In: CLIN 17: Computational linguistics in the Netherlands
44. Medelyan O, Witten IH (2006) Thesaurus based automatic keyphrase indexing. In: Marchionini G, Nelson ML, Marshall CC (eds) ACM/IEEE joint conference on digital libraries. ACM, p 296–297. https://doi.org/10.1145/1141753. 1141819
45. Evans DA, Lefferts RG (1995) CLARIT-TREC experiments. Inf Process Manage 31(3):385–395.https://doi.org/10.1016/0306-4573(94)00054-7
46. Ahmad K, Gillam L, Tostevin L (1999) University of surrey participation in trec8: Weirdness indexing for logical document extrapolation and retrieval (wilder). In: TREC-8: 8th text retrieval conference

47. Church KW, Gale WA (1995) Inverse document frequency (idf): A measure of deviations from Poisson. In: ACL 3rd workshop on very large corpora, 1995. Association for Computational Linguistics. Stroudsburg, PA, p 121–130. https://doi.org/10.1007/978-94-017-2390-9_18
48. Zhang Z, Gao J, Ciravegna F (2016) Jate 2.0: Java automatic term extraction with Apache Solr. In: LREC 2016: Proceedings of the 10th international conference on language resources and evaluation, Portorož, 2016. European Language Resources Association, p 2262–2269
49. Astrakhantsev N (2016) ATR4S: Toolkit with state-of-the-art automatic terms recognition methods in Scala. arXiv preprint, arXiv:1611.07804
50. Sclano F, Velardi P (2007) TermExtractor: a Web application to learn the common terminology of interest groups and research communities. In: TIA 2007: 9th conference on terminology and artificial intelligence, 2007
51. Peñas A, Verdejo F, Gonzalo J (2001) Corpus-based terminology extraction applied to information access. In: Corpus linguistics. p 458–465
52. Frantzi KT, Ananiadou S (1999) The C-Value/NC-Value domain independent method for multi-word term extraction. J Nat Lang Process 6(3):145–179. https://doi.org/10.5715/jnlp.6.3_145
53. Lossio-Ventura JA, Jonquet C, Roche M, Teisseire M (2013) Combining C-Value and keyword extraction methods for biomedical terms extraction. In: International symposium on languages in Biology and Medicine, p 45–49
54. Kozakov L, Park Y, Fin T, Drissi Y, Doganata Y, Cofino T (2004) Glossary extraction and utilization in the information search and delivery system for IBM technical support. IBM Syst J 43(3):546–563. https://doi.org/10.1147/sj.433.0546
55. Bordea G, Buitelaar P, Polajnar T (2013) Domain-independent term extraction through domain modelling. In: TIA 2013: 10th international conference on terminology and artificial intelligence, 2013
56. Badenes-Olmedo C, Redondo-García JL, Corcho O (2017) Efficient clustering from distributions over topics. In: K-CAP 2017: Proceedings of the knowledge capture conference. Austin, TX, 2017. ACM p 17:1–17:8. https://doi.org/10.1145/3148011.3148019
57. Park Y, Byrd RJ, Boguraev B (2002) Automatic glossary extraction: beyond terminology identification. In: 19th international conference on computational linguistics, p 1–7. https://doi.org/10.3115/1072228.1072370
58. Nokel M, Loukachevitch N (2013) An experimental study of term extraction for real information-retrieval thesauri. In: 10th international conference on terminology and artificial intelligence, p 69–76
59. Ermolayev V (2018) OntoElecting requirements for domain ontologies. The case of time domain. EMISA Int J of Conceptual Model. 13(Spl):86–109. https://doi.org/10.18417/emisa.si.hcm.9
60. Kim J-D, Ohta T, Teteisi Y, Tsujii J (2003) GENIA corpus—a semantically annotated corpus for bio-textmining. Bioinformatics 19(suppl. 1):i180–i182. https://doi.org/10.1093/bioinformatics/btg1023
61. Oliver A, V'azquez M (2015) TBXTools: a free, fast and flexible tool for automatic terminology extraction. In: Angelova G, Bontcheva K, Mitkov R (eds) Proceedings of the recent advances in natural language processing. Hissar, Sep 2015, p 473–479
62. Corcho O, Gonzalez R, Badenes-Olmedo C, Dong F (2015) Repository of indexed ROs. Deliverable No. 5.4. Dr Inventor project
63. Gomaa WH, Fahmy AA (2013) A survey of text similarity approaches. Int J Comp Appl 68(13):13–18
64. Yu M, Li G, Deng D, Feng J (2016) String similarity search and join: a survey. Front Comput Sci 10(3):399–417
65. Miller GA, Beckwith R, Fellbaum CD, Gross D, Miller K (1990) WordNet: an online lexical database. Int. J. Lexicograph 3(4):235–244
66. Arnold M, Ohlebusch E (2011) Linear time algorithms for generalizations of the longest common substring problem. Algorithmica 60(4):806–818
67. Levenshtein VI (1966) Binary codes capable of correcting deletions, insertions, and reversals. Sov Phys Dokl 10(8):707–710

68. Hamming RW (1959) Error detecting and error correcting codes. Bell Syst Tech J 29(2):147–160
69. Monger A, Elkan C (1996) The field-matching problem: algorithm and applications. In: Proceedings of the 2nd international conference on knowledge discovery and data mining. AAAI Press, Portland, OR, Aug 1996, p 267–270
70. Jaro MA (1989) Advances in record-linkage methodology as applied to matching the 1985 Census of Tampa Florida. J Amer Stat Assoc 84(406):414–420. https://doi.org/10.1080/01621459.1989.10478785
71. Winkler WE (1990) String comparator metrics and enhanced decision rules in the Fellegi-Sunter model of record linkage. In: Proceedings of the section on survey research, 1990. ASA, p 354–359
72. Dice LR (1945) Measures of the amount of ecologic association between species. Ecology 26(3):297–302
73. Sørensen T (1948) A method of establishing groups of equal amplitude in plant sociology based on similarity of species and its application to analyses of the vegetation on Danish commons. Kongelige Danske Videnskabernes Selskab 5(4):1–34
74. Jaccard P (1912) The distribution of the flora in the alpine zone. New Phytol 11:37–50
75. Huang A (2008) Similarity measures for text document clustering. In: NZCSRSC2008: Proceedings of the 6th New Zealand computer science research student conference, Christchurch, 2008. pp 49–56
76. Singhal A (2001) Modern information retrieval: a brief overview. Bull IEEE Comput Soc Tech Committee Data Eng 24(4):35–43
77. Lu J, Lin C, Wang W, Li C, Wang H (2013) String similarity measures and joins with synonyms. In: Proceedings of the 2013 ACM SIGMOD international conference on the management of data. ACM, p 373–384
78. Lee H, Ng RT, Shim K (2009) Power-law based estimation of set similarity join size. Proc VLDB Endowment 2(1):658–669
79. Tsuruoka Y, McNaught J, Tsujii J, Ananiadou S (2007) Learning string similarity measures for gene/protein name dictionary look-up using logistic regression. Bioinformatics 23(20):2768–2774
80. Qin J, Wang W, Lu Y, Xiao C, Lin X (2011) Efficient exact edit similarity query processing with the asymmetric signature scheme. In: Proceedings of the 2011 ACM SIGMOD international conference on management of data. ACM, New York, NY, p 1033–1044
81. Chowdhury FM, Farrell R (2019) An efficient approach for super and nested term indexing and retrieval. arXiv preprint, arXiv:1905.09761v1 [cs.DS]
82. Karp RM, Rabin MO (1987) Efficient randomized pattern-matching algorithms. IBM J Res Dev 31(2):249–260
83. Knuth DE (1998) The art of computer programming, Volume 3: Sorting and searching, 2nd edn. Addison Wesley Longman Publishing Co., Inc., Redwood City, CA
84. Lu CJ, Browne AC (2012) Development of sub-term mapping tools (STMT). In: AMIA 2012 annual symposium
85. De La Briandais R (1959) File searching using variable length keys. In: IRE-AIEE-ACM'59 (Western): Western joint computer conference. ACM, p 295–298
86. Aho AV, Corasick MJ (1975) Efficient string matching: an aid to bibliographic search. Commun ACM 18(6):333–340
87. Tuarob S, Mitra P, Lee Giles C (2012) Taxonomy-based query-dependent schemes for profile similarity measurement. In: JIWES: 1st joint international workshop on entity-oriented and semantic search at SIGIR 2012, Portland, OR, 2012. ACM, New York, NY

Chapter 3
Formal Framework

In this chapter, the formal framework for the method of detection and measurement of terminological saturation is developed. Terminological saturation is regarded as a phenomenon that occurs in domain-bounded collections of professional texts. The idea of this framework is inspired by the following two factors.

- Data saturation is used by qualitative researchers to have an indication of when interviewees stop bringing new facts that extend the theory, which is being built. If that happens, the theory is considered saturated in interview data and the process is stopped.
- There is a notable similarity in this way of building theories in Qualitative Research with the approach used in the mainstream ontology engineering methodologies (c.f. Sect. 2.6).

Hence, exploiting the phenomenon of saturation in ontology requirements elicitation might be reasonable.

The chapter begins with the preliminaries (Sect. 3.1) that include the definitions helping us frame out the theoretical focus. Next, the research hypotheses that need to be checked with regard to our theoretical framework are presented. These hypotheses are formulated based on the research questions (Sect. 2.13). In Sects. 3.3 and 3.4, the formal introduction of the terminological difference function (*thd*) and the proof of its metric properties in the space of all possible document collections for a subject domain are elaborated. In Sect. 3.5, the theorem shaping the sufficient conditions for the existence of terminological saturation is formulated and proven. Sect. 3.6 elaborates the refinement of the method that is based on the use of document collection partition and corresponding technique of computing merged partial C-Values. This refinement optimizes the method and enables its scalability to industrial size document collections.

V. Kosa and V. Ermolayev, *Terminology Saturation*, Cognitive Science and Technology,
https://doi.org/10.1007/978-981-16-8630-6_3

3.1 Preliminaries

In this work, a set of terms is sought that representatively describe an arbitrary subject domain (*Dom*), for which a domain ontology needs to be developed or refined. Our assumption is that, if we have a reasonably bounded Dom, the set of terms, used to describe it, is finite and not very large.[1] These terms could be extracted from the documents (*Doc*), belonging to a documents collection ($DC = \{Doc\}$) that describes *Dom*. Hypothetically, one may collect all the existing documents that describe *Dom*—a complete document collection.

Definition 3.1 (*A Complete Document Collection for Dom*) A *DC* containing all the documents describing *Dom* is a Complete *DC* (*CDC*).

For any realistic *Dom*, its *CDC* may be very big. Therefore, extracting a representative set of terms from it would be a tedious and resource-consuming task. This is why the document sub-collections of a *CDC* for the *Dom* are examined.

Definition 3.2 (*A document sub-collection*) A document sub-collection (*DSC*) for the *Dom* is the subset of the *CDC*: $DSC \subset CDC$.

Our interest lies in finding a *DSC* of a minimal possible size that contains practically the same set of terms as the *CDC*. The sets of terms are considered practically the same if the terminological difference between the bags of terms extracted from these *DSC* and *CDC* is negligible. Hence, such a *DSC* of minimal possible size describes the *Dom* as fully as the *CDC* for this *Dom*.

Our premise is that, for a reasonably bounded domain, one may observe terminological saturation in the incrementally extended sequence of *DSC*. A terminologically saturated *DSC* will contain practically the same set of terms as the *CDC*.

Several preliminaries have to be defined before explaining how terminological saturation is detected and measured.

A Terminological Basis of a Domain. This notion clarifies what is that a *DC* describes *Dom*. In this work, we understand such a description as containing a subset of valid and significant features that characterize *Dom*. These features could further be used as the requirements for building an ontology for *Dom* and are labelled using the terms, from the documents in a *DC*, describing the domain.

Definition 3.3 (*A terminological basis*) The finite set of terms t_i, labelling all the features that characterize *Dom*, forms the terminological basis $TB = \{t_i\}$, $i = 1, \ldots, dim$ of *Dom*.

Without any loss in generality, a simplified case may be examined, in which all these terms t_i are orthogonal—i.e. are fully different in their senses. More complex cases, in which partially similar terms are considered, could be simplified using a similar terms grouping technique (c.f. [1]), as explained in Sect. 4.4.

---

[1] For example, compared to the total number of words or meaningful phrases in the available documents that describe Dom.

By Definition 3.3, a *TB* contains all the terms relevant for the domain. These terms might not be equally significant for describing the domain, as reflected in the documents, contained in the *CDC*. Let us suppose that a real positive value (*score*) is associated with every term used in the documents of the *CDC*. The more significant the term is for describing the domain, the higher is the *score*. Hence, a vector space model (VSM) could be used as an appropriate formal representation of a document space for *Dom*. In this model, any document or *DC*, including *CDC*, is a point in a vector space with the basis $TB = \{t_i\}$ having dimension *dim*. In this context, it is useful to have a way to compute the position of a document (collection) in the document space and the distance between different documents or collections.

It is not easy, however, to build a *TB* for any realistic domain. A technique is required to extract t_i from the documents describing *Dom*, making sure that an extracted $t_i \in TB$, and that all such t_i have been extracted.

Extracted Terms. Let us suppose now that there is a mapping that transforms a *DC* into a bag of terms (*B*) extracted from the documents of this *DC*. In our work, this mapping is operationalized in the ATE pipeline. Every element *b* in *B* is a pair $\langle t, \text{score}\rangle$, where *t* is a term candidate string and $score$ is an estimate of the likelihood that *t* is a relevant term for the *Dom*: the higher the $score$, the more likely $t \in TB$. The term candidates having high scores are denoted as **significant terms**.

Retained Significant Terms. Let us suppose now that a term significance threshold (*eps*) is rationally estimated for the term scores in a *B*. This threshold indicates a boundary above which the terms are regarded as relevant and valid, hence belong to *TB*. It is also used by many statistical ATE methods to cut off those $b \in B$ that are not likely to be valid terms in *Dom*. After having built a *B*, let us retain the *b*s with the *score* > *eps* in the corresponding bag of retained significant terms (*T*)—c.f. Definition 3.5 below.

In the majority of ATE methods, term significance threshold is either chosen empirically or selected based on common sense considerations. We offer a rationale for the choice of an *eps*.

A Simple Majority Vote. Following [2], a subset of term candidates is considered as containing all significant terms, if the terms in it reflect the sentiment of a simple majority of the knowledge stakeholders in the domain. The premise is that these sentiments are contained in the documents, authored by the domain knowledge stakeholders, describing the domain. The pointers to these sentiments are the terms used to describe them. If a term is used more intensively by the authors, then its influence on the description of the domain is more significant. The more significant the term is the more sentiments of the knowledge stakeholders it indirectly reflects. Therefore, the score of a term might be interpreted as the sum of the **votes** for this term by the domain knowledge stakeholders. A simple majority of voters is reached when the sum of their votes is higher than half of the sum of all votes.

Definition 3.4 (*An individual term significance threshold*) Let

$$B_{desc} = \{b_i = \langle t_i, score_i\rangle\} \tag{3.1}$$

be the bag of extracted term candidates B with the entries sorted in the descending order of their scores. Then an individual term significance threshold for the bag of extracted terms B is computed as follows:

$$eps = score_i : \sum_{j=1}^{i} score_j > \frac{1}{2} \sum_{j=1}^{||B||} score_j, \tag{3.2}$$

where i is the minimal number such that the condition after the colon holds.

Definition 3.4 specifies that, for computing the eps for a B, the minimal subset of bs at the top of the B_{desc} is found, whose voters are the simple majority. Hence, the sum of their scores is higher than ½ of the total sum of the scores in the entire bag of term candidates. The cut-off line below this minimal subset of bs is drawn by setting eps equal to the significance score of the first term[2] below the cut-off line.

Definition 3.5 (*A bag of retained significant terms*) Let $T \subset B_{desc} = \{b_i = \langle t_i, score_i \rangle\}$ for which the following condition (3.3) holds:

$$\forall i, score_i > eps, \tag{3.3}$$

where eps is the individual term significance threshold in B. Then T is the bag of significant terms retained from B.

Successive Approximation for building a TB. If a mapping $DC \rightarrow T$ is available, one (superficial) way for building a *TB* might be to extract and retain all the significant terms from the *CDC* for *Dom*. The terms in the bag of significant terms T_{CDC} will actually constitute the *TB* for *Dom*. This is however hardly tractable for any realistic *Dom* because of at least two reasons:

(i) One has to ensure that the collection of documents she possesses is a *CDC*, which is hard; and
(ii) Processing a *DC*, that qualifies to be a *CDC* for a realistic domain, might be very computationally expensive because of its volume.

A more feasible way is to use a representative *DSC* instead of the *CDC*. Then, the relevant and valid terms extracted from this *DSC* will form a basis that is very similar to *TB*, with a negligible difference. Hence, we need a way to figure out if a *DSC* is representative. One option might be to check if the set of relevant and valid terms, extracted from the *DSC*, is statistically the same as *TB*. This is hard to check using direct comparison, as *TB* is intractable to be built. Therefore, an indirect method has to be employed. In this regard, a technique using successive approximation might be applicable.

[2] This could also be a group of terms having equal scores. If so, a cut-off line is drawn under this group of terms.

Terminological Saturation. Let the sequence of incrementally extended document sub-collections be denoted as $DSC_1, \ldots, DSC_i, \ldots$ Hence, it is true that $DSC_1 \subset \ldots \subset DSC_i \subset \ldots \subset CDC$. Let the bags of retained significant terms extracted from $DSC_1, \ldots, DSC_i, \ldots$ are denoted as $T_1, \ldots, T_i, \ldots$ Let us further suppose that there exists a function (*thd*) that could be used for comparing T_i, T_{i+1} and returning the difference as a real positive value. If at some *i*:

(i) *thd* decreases below the statistical error ε; and
(ii) there is evidence that *thd* does not go above this threshold in the subsequent iterations

then the difference between T_i and T_{CDC} would not be higher than ε. The set of retained significant terms in such a T_i could therefore represent as an ε-approximation of the *TB*. It may also be stated that such a T_i, that could be labelled as T_{sat}, is a **saturated term-set**, B_{sat} is the bag of term candidates from which T_{sat} is retained, and DSC_{sat} is a saturated *DSC* of *Dom*. The difference (*thd*) between T_{sat} and any successive T, including T_{CDC}, is within the admissible and negligible error: $thd(T_{sat}, T_{CDC}) < \varepsilon$.

Terminological Saturation Threshold. The premise for a rational choice of the threshold (ε) for detecting terminological saturation is that a set of terms becomes saturated if it already contains all the terms from *TB*. Hence, whatever terms could be added in subsequent Ts, these are not significant. Let us set $\varepsilon = eps_{B_{sat}}$—the individual term significance threshold computed for B_{sat}. Then T_{sat} will contain statistically the same set of terms as *TB*.

Definition 3.6 (*A terminological core sub-collection*) A terminological core sub-collection (DSC_{sat}) is the sub-collection DSC_i from which the saturated set of significant terms T_{sat} has been retained.

Terminological Saturation Condition. The indicator of terminological saturation for T_i is (c.f. [2]):

$$\begin{aligned} &\text{(i) } thd(T_i T_{i-1}) < eps_i; \text{ and} \\ &\text{(ii) } \forall j > i,\ thd\left(T_j T_{j-1}\right) < eps_j. \end{aligned} \tag{3.4}$$

3.2 Research Hypotheses

Based on the preliminaries shaping out the formal model of terminological saturation (Sect. 3.1), the hypotheses are now put forward that need to be validated for a well-founded formal framework. The hypotheses are formulated in a way to prove the possibility to resolve the issues related to the volatility of *thd* values (c.f. Sect. 5.2.3) causing the possible instability of terminological saturation. These statements form the logic of the developed formal framework.

As it is stated in the terminological saturation condition (3.4), saturation is achieved if particular behaviours of the *thd* and *eps* functions are: (i) observed—the necessary condition; and (ii) proven to persist—the sufficient condition. Hence, the formal properties of the *thd* and *eps* functions need to be studied and the conditions under which the existence of terminological saturation is proven have to be found.

The following hypotheses are put forward, in the above-mentioned context, to be proven in our theoretical framework. If proven, these will answer, at least in part, some of our research questions related to the research objective **RO1** (Table 2.5).

Hypothesis H3.1 (*thd is a metric function*) Let a documents space (*CDS*) be a vector space formed of the VSM representations of all possible individual documents and document sub-collections of a *CDC* in *Dom*. If *thd* is introduced as a kind of a Manhattan distance function [4], then $thd(T_i, T_j)$ is a metric function and *CDS* is a metric space.

For proving this statement, the metric conditions for $thd(T_i, T_j)$ need to be checked: (i) non-negativity; (ii) triangle inequality; (iii) symmetry; and (iv) identity of indiscernibles. Further, it has to be proven that a *CDS* is a metric space with $thd(T_i, T_j)$ as its distance metric.

Hypothesis H3.2 (*Existence of terminological saturation*) Terminological saturation exists if there exist: (i) a monotonically non-decreasing function $eps_{min}(i)$; a monotonically non-increasing function $thds_{max}(i)$; the intersection of these functions at some i.

For terminological saturation measurement, we are interested in computing thd not for arbitrary T_i, T_j, but for successive pairs $T_i, T_{i+1}, i = 1, 2, \ldots$. Hence, the function $thds(i) = thd(T_i, T_{i+1})$ is of our practical interest. In particular, we are interested when $thds(i)$ goes below $eps_{B_{sat}}$. To find this out, we have to analyse:

- The values of individual term significance thresholds eps_i, used for retaining terms in T_i, which could be regarded as a function $eps(i) = eps_i, i = 1, 2, \ldots : B_i \rightarrow \Re^+$.
- The values of $thds(i), i = 1, 2, \ldots : \{T_i, T_{i+1}\} \rightarrow \Re^+$.

As it is revealed in our experiments (Chap. 5), $eps(i)$ is not necessarily a monotonically non-decreasing function. However, it might be possible to build its approximation, $eps_{min}(i)$, as a lower envelope function, that is a monotonically non-decreasing function.

Analogously to $eps(i)$, $thds(i)$ is not necessarily a monotonically non-increasing function. It might also be possible to build its approximation, $thds_{\max}(i)$ as an upper envelope function, that is a monotonically non-increasing function.

Hypothesis H3.3 (*Merged Partial C-Values are equal to conventional C-Values*) Merged partial *C*-Values (MPCV), as denoted by Definition 3.9 (Sect. 3.7), are the sums of the partial *C*-Values (c.f. Definition 3.8, Sect. 3.7) for particular terms along all the partitions of the document collection. The hypothesis is that, for any particular term, its MPCV equals its *C*-Value, where *C*-Values are computed according to [3].

If the hypothesis is valid, it opens the way to compute valid *C*-Values using only the partitions of the collection instead of its incremental parts. This way might be sufficiently more computationally efficient compared to the baseline method.

3.3 Terminological Difference Function (*thd*)

Based on the preliminaries presented in Sect. 3.1, let us now define the function *thd* for measuring the distance between term sets. As inputs, *thd* takes a pair of the bags of terms and returns a real positive value of the distance between these bags of terms. Hence, it is the mapping from the set of all possible pairs of the bags of (retained significant) terms to positive reals:

$$thd : \{\langle T_i, T_j \rangle\} \rightarrow \Re^+ \tag{3.5}$$

While constructing *thd*, it has to be taken into account that, in our approach, the document collections, from which the term-sets are extracted, are incrementally extended such that $DSC_1 \subset DSC_2 \ldots \subset DSC_i \ldots \subset CDC$. Therefore,

$$DSC_{i+1} = DSC_i \cup INC, \tag{3.6}$$

where *INC* is the set of documents, describing *Dom*, which are added to extend the sub-collection for making it potentially more representative.

Due to the incremental nature of sub-collections DSC_i, not only the number of extracted terms will grow in B_{i+k} compared to B_i, but also the scores of the terms. Therefore, for making the scores comparable in the pairs, normalized score values have to be used. Let $maxscore$ be the *score* of the term, having maximal value. In B_{desc} (3.1), *maxscore* is the *score* of the first element. Then a normalized *score* (*ns*) could be computed as:

$$ns = score/maxscore \tag{3.7}$$

Let us now introduce *thd* as a kind of Manhattan distance function.

By definition [4], Manhattan (or also called taxicab) distance is:

$$dman(v_i, v_j) = \sum_{k=1}^{n} \left| v_i^k - v_j^k \right|, \tag{3.8}$$

where $v_i = (v_i^1, v_i^2, \ldots, v_i^n)$, $v_j = \left(v_j^1, v_j^2, \ldots, v_j^n\right)$ are the vectors in an n-dimensional real vector space with fixed cartesian coordinate system.

Following the VSM approach, let us regard $TB = \{t^1, t^2, \ldots, t^n\}$ as a basis for the document's space describing *Dom*. Then, any document sub-collection DSC_i, having the corresponding set of retained significant terms T_i could be mapped to the documents space *CDS* with the basis *TB* as a vector $V_i = (ns_i^1, ns_i^2, \ldots, ns_i^n)$, where ns are normalized significance scores (3.7). T_i, being a set of retained significant terms of the sub-collection DSC_i, may not contain all the terms of the *TB*. Therefore, some of the ns_i in V_i will be zero.

Let DSC_i and DSC_j are different document sub-collections of the same *CDC*. Let also DSC_i and DSC_j be built incrementally (3.6). Then the sets of their retained significant terms are T_i and T_j. Similarly to the bags of extracted term candidates B, from which T are retained, every element b in T_i is a pair $b = \langle t_i^k, ns_i^k \rangle$, where $t_i^k \in TB_i \subseteq TB$ and ns_i^k is its normalized significance score.

Due to the incremental nature of DSC_i and DSC_j on one hand and different individual terms significance thresholds, on the other hand, the following is true with regard to T_i and T_j:

$$int(TB_i, TB_j) = int(TB_j, TB_i) = TB_i \cap TB_j \neq \emptyset,$$

$$dif(TB_i, TB_j) = TB_i \backslash int(TB_i, TB_j) \neq dif(TB_j, TB_i) = TB_j \backslash int(TB_j, TB_i),$$

$$int(TB_i, TB_j) \cap dif(TB_i, TB_j) \cap dif(TB_j, TB_i) = \emptyset \tag{3.9}$$

Hence, the distance between T_i and T_j could be computed as the sum of the three partial distances (3.10)–(3.12):

$$thd_{int}(TB_i, TB_j) = \sum_{k=1}^{\|int(TB_i, TB_j)\|} \left|ns_i^k - ns_j^k\right|, \tag{3.10}$$

where $\|int(TB_i, TB_j)\|$ is the number of elements in $int(TB_i, TB_j)$; ns_i^k, ns_j^k are the individual normalized scores of the same term t^k in T_i and T_j; $t^k \in int(TB_i, TB_j)$.

$$thd_{dif}(TB_i, TB_j) = \sum_{k=1}^{\|dif(TB_i, TB_j)\|} \left|ns_i^k - 0\right| = \sum_{k=1}^{\|dif(TB_i, TB_j)\|} ns_i^k, \tag{3.11}$$

where ns_i^k are the individual normalized scores of the term t^k in T_i; $t^k \in dif(TB_i, TB_j)$.

$$thd_{dif(TB_j,TB_i)} = \sum_{k=1}^{||dif(TB_j,TB_i)||} \left|ns_j^k - 0\right| = \sum_{k=1}^{||dif(TB_j,TB_i)||} ns_j^k, \tag{3.12}$$

where: ns_j^k are the individual normalized scores of the term t^k in T_j; $t^k \in dif(TB_j, TB_i)$.

All the three partial distances (3.10)–(3.12) are Manhattan distances [4] for different subspaces in *CDS*.

Further, the distance between T_i and T_j could be computed as:

$$\begin{aligned} thd(T_i, T_j) &= thd_{int(TB_i,TB_j)} + thd_{dif(TB_i,TB_j)} + thd_{dif(TB_j,TB_i)} \\ &= \sum_{k=1}^{||int(TB_i,TB_j)||} \left|ns_i^k - ns_j^k\right| + \sum_{k=1}^{||dif(TB_i,TB_j)||} ns_i^k + \sum_{k=1}^{||dif(TB_j,TB_i)||} ns_j^k. \end{aligned} \tag{3.13}$$

Hence, $thd(T_i, T_j)$ is also a Manhattan distance in CDS with the basis $TB = \{t^1, t^2, \ldots, t^n\}$. To ensure that $\forall k = 1, \ldots, ||int(TB_i, TB_j)|| t_i^k = t_j^k$, T_i, T_j have to be pre-sorted using their TB_i, TB_j as the key sets for sorting.

Therefore, to compute thd as proposed in (3.13), it is needed to:

- Extract B_i from DSC_i and B_j from DSC_j.
- Compute individual term significance thresholds eps_i for B_i and eps_j for B_j using (3.2).
- Retain significant terms in T_i and T_j.
- Compute ns by applying (3.7) for T_i and T_j.
- Apply (3.13) to compute the distance between T_i and T_j.

3.4 Metric Properties of the *thd* Function

In Sect. 3.3, $thd(T_i, T_j)$ has been introduced as a Manhattan distance function between the sets of retained significant terms T_i and T_j with normalized terms significance scores (*ns*). Therefore, *thd* has all the properties of Manhattan distance [4], which satisfies all Hilbert's axioms for Euclidian spaces with cartesian coordinate basis, except for the side-angle-side axiom.

Let us now prove that thd(T_i, T_j) is a **metric function** for a *CDS*. For that, it is first proven that a *CDS*, with $thd(T_i, T_j)$ as its distance measure, is a space in Lemma 3.1.

Lemma 3.1 *CDS, with* $thd(T_i, T_j)$ *as its distance measure, is a space.*

Proof. As $thd(T_i, T_j)$ is a Manhattan distance, it satisfies (c.f. [4]) all the axioms of a Hilbert space. Therefore, *CDS*, with $thd(T_i, T_j)$ as its distance measure, is a Hilbert space.[3] Hence, *CDS*, with $thd(T_i, T_j)$ as its distance measure, is a space. □

Let us now prove that $thd(T_i, T_j)$ is a metric function in *CDS* in Theorem 3.2.

Theorem 3.2 *$thd(T_i, T_j)$ is a metric function and CDS is a metric space.*

Proof. By the definition of a metric function (c.f. [5]), *thd* has to be a function, measuring the distance between the elements of a set, which satisfies the following conditions:

(i) Non-negativity: $\forall T_i, T_j \in CDS\ thd(T_i, T_j) \geq 0$.
(ii) Triangle inequality: $thd(T_i, T_j) + thd(T_j, T_m) \geq thd(T_i, T_m)$.
(iii) Symmetry: $thd(T_i, T_j) = thd(T_j, T_i)$.
(iv) Identity of indiscernibles: $(thd(T_i, T_j) = 0) \equiv (T_i = T_j)$.

If condition (iv) is not satisfied but replaced by $thd(T_i, T_i) = 0$ then $thd(T_i, T_j)$ is a pseudometric.

(i) **Non-negativity**. $thd(T_i, T_j)$ is a non-negative function as none of the summands in (3.13) could be negative. $\left|ns_i^k - ns_j^k\right|$ is at least equal to 0 (if $ns_i^k = ns_j^k$), or otherwise positive.
(ii) **Triangle inequality**.

$$
\begin{aligned}
& lthd(T_i, T_j) + thd(T_j, T_m) \\
&= \sum_{k=1}^{\|int(TB_i, TB_j)\|} \left|ns_i^k - ns_j^k\right| + + \sum_{k=1}^{\|dif(TB_i, TB_j)\|} ns_i^k + \sum_{k=1}^{\|dif(TB_j, TB_i)\|} ns_j^k \\
&+ \sum_{k=1}^{\|int(TB_j, TB_m)\|} \left|ns_j^k - ns_m^k\right| + \sum_{k=1}^{\|dif(TB_j, TB_m)\|} ns_j^k + \sum_{k=1}^{\|dif(TB_m, TB_j)\|} ns_m^k \\
&= \left(\sum_{k=1}^{\|int(TB_i, TB_j)\|} \left|ns_i^k - ns_j^k\right| + \sum_{k=1}^{\|int(TB_j, TB_m)\|} \left|ns_j^k - ns_m^k\right| \right) \\
&+ \left(\sum_{k=1}^{\|dif(TB_i, TB_j)\|} ns_i^k + \sum_{k=1}^{\|dif(TB_j, TB_m)\|} ns_j^k \right)
\end{aligned}
$$

[3] Strictly speaking, CDS could be regarded as a space only if it contains an infinite number of elements. However, for realistic domains, the number of documents in a CDC, though is finite, could be measured in hundreeds of thousands. Therefore, the number of all possible combinations of these documents and subsets of documents is very big. Hence, considering that a CDC contains an infinite number of elements is a realistic approximation.

$$+\left(\sum_{k=1}^{\|dif(TB_m,TB_j)\|} ns_m^k + \sum_{k=1}^{\|dif(TB_j,TB_i)\|} ns_j^k\right).$$

To prove that *thd* satisfies the triangle inequality condition (ii), it is therefore sufficient to prove that (3.14)–(3.16) all hold true:

$$\left(\sum_{k=1}^{\|int(TB_i,TB_j)\|} \left|ns_i^k - ns_j^k\right| + \sum_{k=1}^{\|int(TB_j,TB_m)\|} \left|ns_j^k - ns_m^k\right|\right) \geq \geq \sum_{k=1}^{\|int(TB_i,TB_m)\|} \left|ns_i^k - ns_m^k\right| \tag{3.14}$$

$$\left(\sum_{k=1}^{\|dif(TB_i,TB_j)\|} ns_i^k + \sum_{k=1}^{\|dif(TB_j,TB_m)\|} ns_j^k\right) \geq \geq \sum_{k=1}^{\|dif(TB_i,TB_m)\|} ns_i^k \tag{3.15}$$

$$\left(\sum_{k=1}^{\|dif(TB_m,TB_j)\|} ns_m^k + \sum_{k=1}^{\|dif(TB_j,TB_i)\|} ns_j^k\right) \geq \geq \sum_{k=1}^{\|dif(TB_m,TB_i)\|} ns_m^k \tag{3.16}$$

Let us recall that $TB_i \subseteq TB, TB_j \subseteq TB, TB_m \subseteq TB$ due to the incremental design of document sub-collections DSC_i. Hence, without any loss of distance value, (3.14)–(3.16) could be transformed as shown in (3.17)–(3.19) below. The left part of (3.14) could be rewritten as follows:

$$\begin{aligned}
&\sum_{k=1}^{\|int(TB_i,TB_j)\|} \left|ns_i^k - ns_j^k\right| + \\
&+ \sum_{k=1}^{\|int(TB_j,TB_m)\|} \left|ns_j^k - ns_m^k\right| = \\
&= \sum_{k=1}^{\|TB\|} \left|ns_i^k - ns_j^k\right| + \sum_{k=1}^{\|TB\|} \left|ns_j^k - ns_m^k\right| = \\
&= \sum_{k=1}^{\|TB\|} \left(\left|ns_i^k - ns_j^k\right| + \left|ns_j^k - ns_m^k\right|\right),
\end{aligned} \tag{3.17}$$

where: ns_i^k, ns_j^k, ns_m^k correspond to those $t^k \in TB$ which also belong to TB_i, TB_j, TB_m; and $\left(\left|ns_i^k - ns_j^k\right| + \left|ns_j^k - ns_m^k\right|\right) \geq \left|ns_i^k - ns_m^k\right|$ as triangle inequality holds for distances between real numbers. This proves (3.14).

Similarly, the left part of (3.15) could be rewritten as follows:

$$\left(\sum_{k=1}^{\|dif(TB_i,TB_j)\|} ns_i^k + \sum_{k=1}^{\|dif(TB_j,TB_m)\|} ns_j^k\right) = \\ = \sum_{k=1}^{\|TB\|} \left(ns_i^k + ns_j^k\right) \geq \sum_{k=1}^{\|TB\|} ns_i^k, \tag{3.18}$$

as any $ns_j^k > eps > 0$. This proves (3.15).

Consequently, rewriting the left part of (3.16) analogously to (3.18) we have:

$$\left(\sum_{k=1}^{\|dif(TB_m,TB_j)\|} ns_m^k + \sum_{k=1}^{\|dif(TB_j,TB_i)\|} ns_j^k\right) = \\ = \sum_{k=1}^{\|dif(TB)\|} \left(ns_m^k + ns_j^k\right) \geq \sum_{k=1}^{\|TB\|} ns_m^k \tag{3.19}$$

This proves (3.16) and finally proves that triangle inequality holds for *thd*.

(iii) **Symmetry**. The symmetry of *thd* is proven, using (3.13) and the fact that the union operation for sets is symmetric, i.e. $int(TB_i, TB_j) = TB_i \cap TB_j = TB_j \cap TB_i = int(TB_j, TB_i)$:

$$thd(T_i, T_j) = \\ = \sum_{k=1}^{\|int(TB_i,TB_j)\|} \left|ns_i^k - ns_j^k\right| + \sum_{k=1}^{\|dif(TB_i,TB_j)\|} ns_i^k + \\ + \sum_{k=1}^{\|dif(TB_j,TB_i)\|} ns_j^k = \\ = \sum_{k=1}^{\|int(TB_j,TB_i)\|} \left|ns_j^k - ns_i^k\right| + \sum_{k=1}^{\|dif(TB_j,TB_i)\|} ns_j^k + \\ + \sum_{k=1}^{\|dif(TB_i,TB_j)\|} ns_i^k = \\ = thd(T_j, T_i)$$

(iv) **Identity of indiscernibles**. It implies from (3.13) that:

$$\begin{aligned} &thd(T_i, T_i) = \\ &= thd_{int(TB_i,TB_i)} + thd_{dif(TB_i,TB_i)} + thd_{dif(TB_i,TB_i)} = \\ &= \sum_{k=1}^{\|int(TB_i,TB_i)\|} \left|ns_i^k - ns_i^k\right| + 0 + 0 = 0 + 0 + 0 = \\ &= 0. \end{aligned} \tag{3.20}$$

Therefore, $thd(T_i, T_i) = 0$.

On the other hand, according to (3.13), $\left(thd\left(T_i, T_j\right) = 0\right)$ implies that

$$\begin{aligned} &\left(\sum_{k=1}^{\|int(TB_i,TB_j)\|} \left|ns_i^k - ns_j^k\right| = 0\right) \wedge \\ &\wedge \left(\sum_{k=1}^{\|dif(TB_i,TB_j)\|} ns_i^k = 0\right) \wedge \\ &\wedge \left(\sum_{k=1}^{\|dif(TB_j,TB_i)\|} ns_j^k = 0\right). \end{aligned}$$

$\sum_{k=1}^{\|dif(TB_i,TB_j)\|} ns_i^k = 0$ means that $dif\left(TB_i, TB_j\right) = \emptyset$ as any $ns_i^k > eps_i > 0$. The same is true for $dif\left(TB_i, TB_j\right)$. Therefore, $TB_i = TB_j = int\left(TB_i, TB_j\right)$, which means that T_i and T_j contain the same set of terms.

$\sum_{k=1}^{\|int(TB_i,TB_j)\|} \left|ns_i^k - ns_j^k\right| = 0$ means that $\forall t^k \in int\left(TB_i, TB_j\right) ns_i^k = ns_j^k$. This may only happen if $i = j$ as different T_i and T_j are formed using the documents from different DSC_i and DSC_j. Even if the documents in these collections are all the same[4] (hence, the terms in T_i and T_j are all the same), the significance values of the terms are different.

Therefore, $\left(thd\left(T_i, T_j\right) = 0\right) \rightarrow \left(T_i = T_j\right)$ as $i = j$ and $T_i = T_i$ by the definition of T.

Thus, $thd\left(T_i, T_j\right)$ is a metric function in *CDS*. Consequently, *CDS* is a metric space, as $thd\left(T_i, T_j\right)$ is a metric distance function in it. □

The proof of Theorem 3.2 formally validates our research hypothesis **H3.1** (Sect. 3.2).

[4] This could be true if documents are replicated to form synthetic sub-collections, as for example was done in [6].

3.5 Existence Conditions for Terminological Saturation

Let us now prove our hypothesis **H3.2**, which is about the sufficient condition for the existence of terminological saturation. For that, the hypothesis is formulated as the existence theorem.

Theorem 3.3 (Sufficient conditions of terminological saturation) *Let:*

(i) $DSC_1 \subset DSC_2 \subset \ldots \subset DSC_i \ldots$ *be the sequence of incrementally extended document sub-collections, each describing the same arbitrary domain Dom.*

(ii) $B_1, B_2, \ldots, B_i, \ldots$ *be the sequence of the bags of term candidates extracted from* $DSC_1, DSC_2, \ldots, DSC_i \ldots$ *and* $eps(i)$ *is the function of individual term significance thresholds for* $B_i, i = 1, 2, \ldots$

(iii) $T_1, T_2, \ldots, T_i, T_{i+1}, \ldots$ *be the sequence of the bags of retained significant terms for which pairwise successive terminological difference is computed using the* $thds(i)$ *function.*

Then, the sequence of DSC_i *is terminologically saturated,* i *is the saturation point, and* $TB_i = TB_{sat} = TB$*, if:*

(i) *There exist a non-decreasing lower envelope function* $eps_{min}(i)$ *for* $eps(i)$.

(ii) *There exist a non-increasing upper envelope function* $thds_{max}(i)$ *for* $thds(i)$.

(iii) $eps_{min}(i) \geq thds_{max}(i)$.

Proof. Let us suppose that $eps(i)$ and $thds(i)$ are available—i.e. their values for $i = 1, 2, \ldots$ could be computed. Based on that, let us build the required envelope functions $eps_{min}(i)$ and $thds_{max}(i)$.

Let $eps_{min}(1) = eps(1)$. Then $eps_{min}(i)$ will be non-decreasing iff:

$$\forall i > 1 eps(i) \geq eps_{min}(1) \tag{3.21}$$

Let us suppose that $\exists j > 1 : \forall i > 1, eps(j) = \min(eps(i))$. Then $eps_{min}(j) = eps(j)$ and

$$\begin{aligned} \forall i = 2, \ldots, j-1, eps_{min}(i) = eps_{min}(1) + \\ +(i-1)\frac{eps_{min}(j) - eps_{min}(1)}{j-1} \end{aligned} \tag{3.22}$$

Let us further suppose that $eps_{min}(k)$ has been computed. Then, the previous iteration is repeated as follows:

The minimal $j > k$ is sought such that $\forall i > k \; eps(j) = min(eps(i))$. Then $eps_{min}(j) = eps(j)$ and the values of $eps_{min}(i)$ for $i = k+1, \ldots, j-1$ are computed as:

$$\forall i = k+1, \ldots, j-1$$

$$eps_{min}(i) = eps_{min}(k) + \\ +(i-k)\frac{eps_{min}(j) - eps_{min}(k)}{j-k} \quad (3.23)$$

Let us now build $thds_{max}(i)$ in a way that is similar to what has been done for building $eps_{min}(i)$. First, let $thds_{max}(1) = thds(1)$. Hence, $thds_{max}(i)$ will be non-increasing iff:

$$\forall i > 1, thds(i) \leq thds_{max}(1) \quad (3.24)$$

Let us further suppose that $thds_{max}(k)$ for some $k = 1, \ldots$ has been computed. The minimal $j > k$ is sought such that $\forall i > k; thds(j) = max(thds(i))$.

Then $thds_{max}(j) = thds(j)$ and the values of $thds_{max}(i)$ for $i = k+1, \ldots, j-1$ are computed as:

$$\forall i = k+1, \ldots, j-1$$

$$thds_{max}(i) = thds_{max}(k) + \\ +(i-k)\frac{thds_{max}(j) - thds_{max}(k)}{j-k} \quad (3.25)$$

Therefore, the existence of $eps_{min}(i)$ and $thds_{max}(i)$ is proven by constructing the functions under the conditions (3.21) and (3.24) respectively. Hence, the conditions (i) and (ii) of the theorem are met if (3.21) and (3.24) hold.

Let us now suppose that $\exists i > 1 : eps_{min}(i) \geq thds_{max}(i)$. Then,

$$\forall j > i, eps_{min}(j) \geq eps_{min}(i) \geq \\ \geq thds_{max}(i) \geq thds_{max}(j) \quad (3.26)$$

due to (3.23) and (3.25). Hence, condition (iii) of the theorem holds, $eps_{min}(i) = \varepsilon = eps_{B_{sat}}$ is the threshold for terminological saturation, i is the saturation point, and $TB_i = TB_{sat} = TB$. □

Theorem 3.3 provides the approach to detect terminological saturation if it can be reached, along the sequence of $DSC_1 \subset DSC_2 \ldots \subset DSC_i \ldots$ However, it does not help prove that terminological saturation in a particular sequence of document sub-collections is indeed reachable before all the sub-collections are processed. Therefore, it might be valuable to find out if terminological saturation could be predicted based on the small number of initial iterations of the successive approximation process. This research line is planned for future work.

3.6 Scalability and Optimization

The C-Value method [3], chosen in our work for term extraction (Sect. 2.8), is hybrid as it uses both the measures of "unithood" and "termhood" in one ranking formula (3.28). The method pipelines linguistic and statistical phases that are consequently applied to the entire text—a (sub-)collection of documents in our case. In the beginning, the linguistic phase outputs the bag of term candidate strings. After that, the statistical phase computes significance scores for these term candidate strings as C-Values. For us, to process the incrementally extended sequence of texts, this (conventional) C-Value pipeline appears to be too computationally expensive, as demonstrated by runtime diagram in Fig. 5.29 (Chap. 5). This section reports[5] how could the C-Value method by Frantzi and Ananiadou [3] be refined with an aim to reduce, significantly, the incurred computational expense.

Let us now consider a document collection DC as a composition of its disjoint parts.

Definition 3.7 (*A partial collection and a partition of a document collection*) $DC_i, i = 1, \ldots, n$ are the partial document collections of DC and $\{DC_i\} = \{DC_i\}_{i=1}^{n}$ is the partition of DC if the following conditions hold:

$$\text{Condition 1}: \quad DC = \bigcup_{i=1}^{n} DC_i$$

$$\text{Condition 2}: \quad \bigcap_{i=1}^{n} DC_i = \emptyset, \qquad (3.27)$$

The linguistic part of the C-Value method processes separate sentences. Therefore: (i) its computational complexity is the function of the number of sentences in a document collection; and (ii) the partial collections $DC_i, i = 1, \ldots, n$ (Definition 3.7) could be processed independently and the outputs further merged. Hence, applying the linguistic step to the partition of DC could at least be parallelized, which results in the runtime gain of n times.

In our pipeline, the linguistic step is iteratively applied to the incrementally enlarged datasets (see Sect. 3.1). Therefore, the same chunks of text are processed many times. Let us suppose that a DC contains k documents and $inc = k/n$ is the increment to enlarge datasets. Then, the number of documents to be processed is:

- In the case of the incrementally enlarged datasets: $inc + 2\ inc + \cdots + n\ inc = k\ \frac{1+2+\cdots+n}{n} \approx (n+1)/2\ k$, which is substantially more than k if $n > 1$.
- In the case of partial collections: k.

Hence, processing partial collections instead of incrementally enlarged datasets give a substantial gain in runtime, which is $\left(\frac{n+1}{2} - 1\right)k$.

[5] The material of this section is based on our publication [7].

In the statistical part of the method, a *C*-Value [3] rank, further labelled as cv in formulae and equations, is computed for every term candidate string provided by the linguistic part of the pipeline. *C*-Value formula (3.28) uses several statistical characteristics of the corresponding term candidate string. These characteristics are:

- The total number of occurrence(s) of the candidate string in the document corpus.
- The number of occurrence(s) of the candidate string as a part of other longer candidate terms.
- The number of these longer candidate terms.
- The length of the candidate string (in the number of words).

Let: s be a term candidate string; $|s|$—the length of s in words; ls—a longer term candidate string in which s is nested as a substring; $f(.)$—the number of occurrence(s) of a term candidate string in a DC; T^s—the set of extracted term candidate strings ls that nest s; and $P(T^s)$—the number of these ls. Then a (complete) cv of s is denoted [3] as follows:

$$cv(s) = \begin{cases} log_2(|s|)\ f(s) & \text{if } s \text{ is not nested in any ls extracted from } DC \\ log_2(|s|)\left(f(s) - \frac{1}{P(T^s)} \sum\limits_{ls \in T^s} f(ls)\right) & \text{otherwise} \end{cases} \quad (3.28)$$

Hence, the computational complexity of $cv(s)$ depends on $P(T^s)$ which might grow substantially with the growth of the processed text dataset. Therefore, it might be reasonable to apply the statistical step of the pipeline not to the incrementally enlarged datasets, but to partial collections.

It is not straightforward, however, that: (i) computing *C*-values for the terms extracted from the partial collections; and (ii) merging further these bags of terms with their significance scores; will give the same result as applying the statistical step to incrementally enlarged datasets. In the remainder of this section, it is proven formally that partitioning *C*-Value computation with further merging of the outputs gives correct results up to the validity of the $h1$ hypothesis in (3.34). This hypothesis is further validated in our experimental study reported in Sect. 5.7.

Definition 3.8 (*Partial C-Value*) The partial *C*-Value of the term candidate string s extracted from the partial document collection DC_i is computed as:

$$pcv_i(s) = \begin{cases} log_2(|s|)\ f_i(s) & \text{if } s \text{ is not nested in any } ls \text{ extracted from } DC_i \\ log_2(|s|)\left(f_i(s) - \frac{1}{P(T_i^s)} \sum\limits_{ls \in T_i^s} f_i(ls)\right) & \text{otherwise} \end{cases}, \quad (3.29)$$

where T_i^s is the set of term candidate strings ls, that nest s, extracted from DC_i; $f_i(.)$ is the number of occurrences of s or ls in DC_i.

Lemma 3.4 (The total number of nested occurrences) *The total value of the number of nested occurrences, in DC, of a term candidate string s in longer-term candidate strings ls is the sum of the total numbers of the nested occurrences of s in all partial collections DC_i of DC*:

$$tnest(s) = \sum_{ls \in T^s} f(ls) = \sum_{i=1}^{n} \left(\sum_{ls \in T_i^s} f_i(ls) \right) = \sum_{i=1}^{n} (tnest_i(s)). \tag{3.30}$$

Proof. It implies from Definition 3.8 (of partial *C*-Value), that $tnest_i(s)$ is the total number of occurrences of the term candidate string *s* in all longer term candidate strings ls extracted from the partial collection DC_i. The number of these longer term candidate strings equals to $P\left(T_i^s\right)$. Due to the disjointness of the partial collections DC_i (Condition 2 of Definition 3.7 (3.27)), $f(ls) = \sum_{i=1}^{n} f_i(ls)$. Therefore, and due to the Condition 1 of Definition 3.7 (3.27), the total number of occurrences of *s* in all *ls* extracted from *DC* is:

$$tnest(s) = \sum_{ls \in T^s} f(ls) = \sum_{ls \in T^s} \sum_{i=1}^{n} f_i(ls) = \sum_{ls \in \bigcup_{i=1}^{n} (T_i^s)} \left(\sum_{i=1}^{n} f_i(ls) \right) =$$

$$= \sum_{i=1}^{n} \left(\sum_{ls \in T_i^s} f_i(ls) \right) = \sum_{i=1}^{n} (tnest_i(s)). \quad \square$$

Definition 3.9 (*Merged partial C-Value*) The merged partial *C*-Value of the term candidate string *s* is computed as:

$$mpcv(s) = \sum_{i=1}^{n} pcv_i(s). \tag{3.31}$$

Theorem 3.5 allows computing $cv(s)$ for the whole collection *DC* based on the merging of the known partial *C*-Values $pcv_i(s), i = 1, \ldots, n$ for the partial collections $DC_i, i = 1, \ldots, n$ of *DC*.

Theorem 3.5 (Equality of *cv* and *mpcv*) *If a document collection DC is partitioned as* $\{DC_i\} = \{DC_i\}_{i=1}^{n}$, *which means that Condition 1 and Condition 2* (3.27) hold, then

$$cv(s) = mpcv(s) \tag{3.32}$$

Proof. The proof is structured in three cases: (i) *s* is never nested in *ls*; (ii) $\forall DC_i$, *s* is nested at least once and at least in one *ls*; and (iii) *s* is nested in *ls* in some DC_i.

Case (i): *not nested*. If, $\forall i = 1, \ldots, n$, *s* extracted from DC_i is not nested in any *ls* extracted from DC_i, then *s* is not nested in any *ls* extracted from DC. Therefore, for such an s:

$$mpcv(s) = \sum_{i=1}^{n} pcv_i(s) = \sum_{i=1}^{n} log_2(|s|)\ f_i(s) =$$
$$= log_2(|s|) \sum_{i=1}^{n} f_i(s) = \log_2(|s|) f(s) = cv(s). \qquad (3.33)$$

due to Condition 1 and Condition 2 (3.27) and the definition of $f(.)$.

Case (ii): *all nested*. If, $\forall i = 1, \ldots, n$, s extracted from DC_j is nested in an ls extracted from DC_j, then: (a) this s (extracted from DC) is nested in this ls (extracted from DC); and (b) $ls \in T_j^s \subset T^s$—because $DC_i \subset DC$ due to Condition 1 (3.27). Therefore:

$$mpcv(s) = \sum_{i=1}^{n} pcv_i(s) =$$
$$= \sum_{i=1}^{n} \left(log_2(|s|) \left(f_i(s) - \frac{1}{P(T_i^s)} \sum_{ls \in T_i^s} f_i(ls) \right) \right) =$$
$$= log_2(|s|) \sum_{i=1}^{n} \left(f_i(s) - \frac{1}{P(T_i^s)} \sum_{ls \in T_i^s} f_i(ls) \right) =$$
$$= log_2(|s|) \left(\sum_{i=1}^{n} f_i(s) - \sum_{i=1}^{n} \left(\frac{1}{P(T_i^s)} \sum_{ls \in T_i^s} f_i(ls) \right) \right) =$$
$$= log_2(|s|) \left(f(s) - \sum_{i=1}^{n} \left(\frac{1}{P(T_i^s)} \sum_{ls \in T_i^s} f_i(ls) \right) \right) \approx_{|h1}$$
$$\approx_{|h1} log_2(|s|) \left(f(s) - \frac{1}{P(T^s)} \sum_{i=1}^{n} \left(\sum_{ls \in T_i^s} f_i(ls) \right) \right) =$$
$$= log_2(|s|) \left(f(s) - \frac{1}{P(T^s)} \sum_{ls \in T^s} (f(ls)) \right) = cv(s). \qquad (3.34)$$

Here "$\approx_{|h1}$" stands for hypothetically approximately equal. The hypothesis $h1$ is about the approximate equality in $\sum_{i=1}^{n} \left(\frac{1}{P(T_i^s)} \right) \approx \frac{1}{P(T^s)}$. Formally, $\sum_{i=1}^{n} \left(\frac{1}{P(T_i^s)} \right) > \frac{1}{P(T^s)}$. However, asymptotically, $P(T_i^s) = o(tnest_i(s))$ and $P(T^s) = o(tnest(s))$ due to: (a) the overlaps in T_i^s; and (b) possible nestings in several instances of ls. Therefore, the influence of those denominators in (3.34

becomes lower with the growth of the volume of DC and its partial collections DC_i. This is a promise that $h1$ might hold true.

Case (iii): s is *sometimes nested* in ls. There exist several partial collections DC_j, for which Case (ii) is applied. For the rest of partial collections DC_k Case (i) is applied. In this situation two partial sums—$mpcv_1(s)$ and $mpcv_2(s)$—are computed for these disjoint subsets of the partition of DC. Similarly to Case (ii), $cv(s) \approx_{|h1} mpcv_1(s) + mpcv_2(s)$.

Hence, if the hypotheses $h1$ holds true, Cases (i)–(iii) prove Theorem 3.5. □

For checking $h1$, complete (*cv*) and merged partial (*mpcv*) *C*-Values are experimentally computed and compared, as presented in Sect. 5.7.

A straightforward corollary from Theorem 3.5 is that *C*-Values do not depend on the partitioning of a document collection.

Corollary 3.6 (Size of a partial collection) *Let* $\{DC_i\}_{i=1}^{n}$; $\left\{DC_j\right\}_{j=1}^{m}$; $n \neq m$ *be two different partitions of a document collection DC. Then:*

$$\forall s,\ mpcv(s)|_{\{DC_i\}} = mpcv(s)|_{\{DC_j\}} \approx_{|h1} cv(s), \tag{3.35}$$

where s is a term candidate string extracted from the document collection DC; $mpcv(s)|_{\{DC_i\}}$ *is the merged partial C-Value of the term candidate string s computed for the partition*$\{DC_i\}$ *of DC;* $mpcv(s)|_{\{DC_j\}}$ *is the merged partial C-Value of the term candidate string s computed for the partition* $\{DC_j\}$ *of DC.*

Based on Corollary 3.6, the size of a partial collection $DC_i \in \{DC_i\}$ may be reasonably chosen based on the specifics of the problem and available hardware resources, in particular—RAM. One example of a possible scenario and problem might be extracting terms from a stream of textual documents, like blog posts or tweets. In this setting, the size of a partial collection has to be smaller than the size of the stream window.

3.7 Summary

In this chapter, we developed a formal framework for terminological saturation measurement in reply to our research questions about:

- **RQ4**: the rational choice of the cut-off threshold (eps) for distinguishing between significant and insignificant term candidate strings.
- **RQ5**: the way to compute terminological saturation between two bags of extracted terms, where every term is provided with its *C*-Value.
- **RQ6**: the sufficient conditions for terminological saturation to exist.
- **RQ7**: the way to optimize the process of the computation of terminological saturation measurement.
- **RQ2**: the optimal size of a data increment.

The questions, in fact, dealt with the three important ingredients of the formal framework: (i) the function for measuring terminological difference (thd) and its formal properties; (ii) the sufficient existence conditions for terminological saturation; and (iii) the formal approach for optimizing the computation in the process of extracting terms. Therefore, to answer the research questions in a formally clear and structured way, the research hypotheses, **H3.1–H3.3**, were formulated and further tested as valid in the chapter.

H.3.1 has been formally tested in Sects. 3.3 and 3.4. The function for measuring terminological difference (thd) between the bags of terms has been formally defined in Sect. 3.3 as a kind of the Manhattan distance function. The definition of thd has indirectly used the definition of the individual term significance threshold (eps) that had been introduced in Sect. 3.1. The metric properties of the thd function have been validated by the proofs of Lemma 3.1 and Theorem 3.2 in Sect. 3.4. Hence, by proving the validity of **H.3.1**, we have answered **RQ4** and **RQ5**.

H3.2 has been formally validated in Sect. 3.5 by the proof of the existence Theorem 3.3. Hence, we have answered **RQ6**. This answer, however, has not allowed computing the predictions of saturation points in arbitrary document collections, based on a few initial iterations. Predictions are beyond the scope of this project and are left for future work, to be approached as proposed in [8].

H3.3 has been formally tested in Sect. 3.6. For testing the hypothesis, the notions of a partition of a document collection, a partial C-Value, and a merged partial C-Value have been introduced. Further, Lemma 3.3 and Theorem 3.4 have formally proven that conventional C-Values are equivalent to merged partial C-Values up to the validity of the $h1$ hypothesis[6] about the approximate equality in $\sum_{i=1}^{n}\left(\frac{1}{P(T_i^s)}\right) \approx \frac{1}{P(T^s)}$. Hence, a positive constructive answer for **RQ7** has been given. The **RQ2** question about an optimal size for dataset increments has also been answered by the proof of Corollary 3.6 to Theorem 3.5. It stated that merged partial C-Values do not depend on the size of such an increment (partition). Therefore, the size of a partition may be reasonably chosen based on the specifics of the problem and available hardware resources (e.g. RAM). This result, in fact, removed the constraints of the software implementations of the C-Value method and allowed our MPCV method to scale efficiently up to the realistic industrial sizes of document collections.

As a summary, it could be stated that all the research questions, put upfront for this chapter have been positively answered by proving the validity of the three research hypotheses. This allowed us to elaborate on all the required components of the formal framework.

---

[6] This hypothesis will be validated experimentally in Sect. 5.7, which will finalise the validity check of **H3.3**.

References

1. Chugunenko A, Kosa V, Popov R, Chaves-Fraga D, Ermolayev V (2018) Refining terminological saturation using string similarity measures. In: Ermolayev V et al (eds) ICTERI 2018: proceedings of the 14th international conference on ICT in education, research, and industrial applications. Volume I: main conference, Kyiv, May 2018. CEUR-WS, vol 2105, pp 3–18
2. Tatarintseva O, Ermolayev V, Keller B, Matzke W-E (2013) Quantifying ontology fitness in OntoElect using saturation- and vote-based metrics. In: Ermolayev V et al (eds) Revised selected papers of ICTERI 2013. Communications in computer and information science, vol 412. Springer, Heidelberg, pp 136–162. https://doi.org/10.1007/978-3-319-03998-5_8
3. Frantzi KT, Ananiadou S (1999) The C-Value/NC-Value domain independent method for multi-word term extraction. J Nat Lang Process 6(3):145–179. https://doi.org/10.5715/jnlp.6.3_145
4. Gomaa WH, Fahmy AA (2013) A survey of text similarity approaches. Int J Comp Appl 68(13):13–18
5. Moore RE, Cloud MJ (2007) Computational functional analysis, 2nd edn. Woodhead Publishing Limited
6. Kosa V, Chaves-Fraga D, Naumenko D, Yuschenko E, Badenes-Olmedo C, Ermolayev V, Birukou A (2018) Cross-evaluation of automated term extraction tools by measuring terminological saturation. In: Bassiliades N et al (eds) Revised selected papers of ICTERI 2017. Communications in computer and information science, vol 826. Springer, Cham, pp 135–163. https://doi.org/10.1007/978-3-319-76168-8_7
7. Kosa V, Chaves-Fraga D, Dobrovolskiy H, Ermolayev V (2020) Optimized term extraction method based on computing merged partial *C*-values. In: Ermolayev V et al (eds) Revised selected papers of ICTERI 2019. Communications in computer and information science, vol 1175. Springer, Cham, pp 24–49. https://doi.org/10.1007/978-3-030-39459-2_2
8. Kosa V, Ermolayev V (2020) Toward a theoretical framework of terminological saturation for ontology learning from texts. In: MS-AMLV 2019: proceedings of the 1st masters symposium on advances in data mining, machine learning, and computer vision, Lviv, November 2019. CEUR-WS, vol 2566, pp 40–51

Chapter 4
Algorithmic Suite

In this chapter, new algorithms are developed or the formerly developed algorithms are improved with an aim to operationalize our formal framework (Chap. 3) for computation.

The chapter begins with the proposal of the processing pipeline in Sect. 4.1. For that, the workflow for detecting and measuring terminological saturation (Fig. 4.1) is specified. The workflow is further detailed in the functional block diagram that reveals the modular structure of our algorithmic suite and dependencies between its modules. The algorithms representing the modules are further elaborated as follows. Section 4.2 presents the algorithms developed to instrument data preparation for further processing. These include the generation of a document collection catalogue and download of the full-text documents (in PDF) based on the catalogued metadata. Section 4.3 presents the pre-processing phase, comprising the algorithms for PDF to plain text conversion, configuration of the computation flow, and generation of the datasets from plain text documents. The optimized algorithm for the statistical part of the term extraction pipeline and the algorithm for merging partial *C*-Values are developed in Sect. 4.4. The baseline algorithm for computing terminological difference between two bags of terms is presented in Sect. 4.5. Section 4.6 refines this baseline algorithm by incorporating the developed term grouping technique, including the algorithms for textual string similarity measurement and the refined algorithm for measuring terminological difference. The algorithm for removing regular accumulated noise from the bags of terms is reported in Sect. 4.7. Section 4.8 outlines the implementation of the algorithmic suite in software and offers the pointers to this software, thus making the results publicly assessable and reproducible. Finally, Sect. 4.9 presents the summary for the chapter.

V. Kosa and V. Ermolayev, *Terminology Saturation*, Cognitive Science and Technology,
https://doi.org/10.1007/978-981-16-8630-6_4

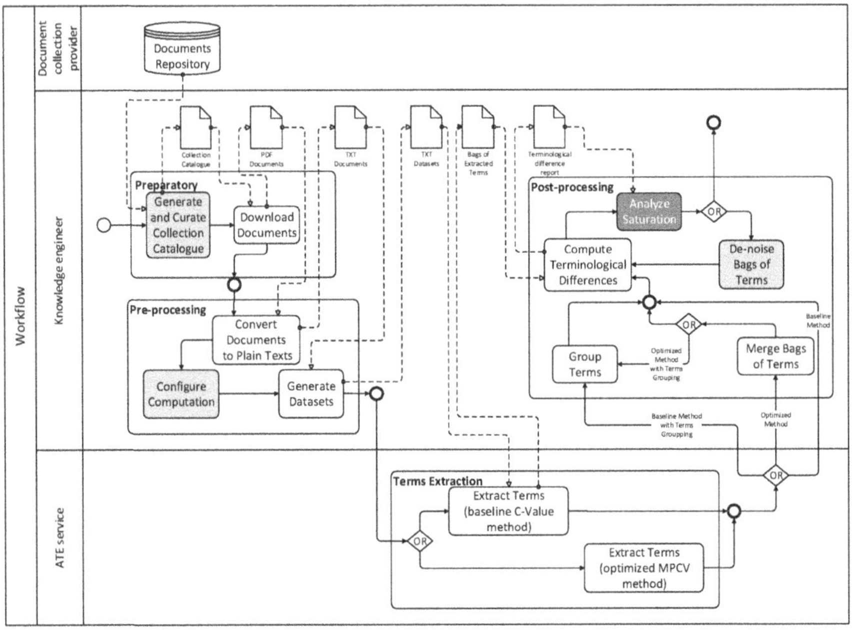

Legend: White-coloured activities are fully automated, light-greyed activities require user interaction, dark-greyed activities are manual.

Fig. 4.1 The computation workflow for terminological saturation detection and measurement

4.1 Computation Flow

As specified in our formal framework for terminology saturation detection and measurement (c.f. Chap. 3), the objective of our method is to arrive at a terminological core document sub-collection of a bigger domain-bounded document collection. If terminological saturation is achieved, the terminological core sub-collection contains the set of terms, which is negligibly different from the set of terms that could have been extracted from the entire collection. For that, the process of iterative approximation towards a terminological core term-set has been formally elaborated in Chap. 3. Every iteration (i) in this process is based on measuring terminological difference ($thd(T_i, T_{i+1})$) between the two bags of retained significant terms (T_i and T_{i+1}) extracted from the two document sub-collections, DSC_i and DSC_{i+1}, where $DSC_{i+1} = DSC_i \cup INC$ (3.6) and INC is the incremental set of documents, which are added to DSC_i to extend this sub-collection for making it potentially more representative. The following steps were proposed in Sect. 3.3 to measure terminological difference at an iteration $= 0, \ldots$:

- Form document sub-collection DSC_{i+1}.
- Extract the bag of terms B_{i+1} from DSC_{i+1} (B_i has been extracted from DSC_i in the previous iteration $(i - 1)$.
- Compute individual term significance threshold eps_{i+1} for B_{i+1}.
- Retain significant terms in T_{i+1}.
- Compute ns for T_{i+1}.
- Compute the terminological distance $thd(T_i, T_{i+1})$.

In this chapter, we aim at operationalizing this successive approximation process by elaborating the workflow, presented in Fig. 4.1 using Business Process Model and Notation (BPMN)[1] as a formal meta-model.

Structurally, the workflow contains the following four phases: (i) Preparatory; (ii) Pre-processing; (iii) Terms extraction; and (iv) Post-processing. The objectives of the phases are:

- **Preparatory**: *assemble the collection* from which the document core sub-collection will be extracted. This includes document sampling, the elaboration of the collection catalogue, and the retrieval of the documents in full texts.
- **Pre-processing**: *prepare the data* for computation. This includes the conversion of the retrieved documents to plain texts, the configuration of the computation process, and the generation of the datasets.
- **Terms extraction**: *extract the bags of terms* from the datasets. In our approach, the extraction could be done by any method, implemented in software, that returns the bag of term candidates (character strings) with their significance values (real numbers). The method is not supposed to do cut-offs of insignificant term candidates. Two different implementations of the *C*-Value method are used in our work (baseline and optimized). The choice of the baseline method is argued in Sects. 2.8, 2.9, and 5.4. The optimized method (MPCV) has been developed in Sect. 3.6. The use of the Aho-Corasick textual strings matching algorithm for optimizing partial *C*-Values computation is argued in Sect. 2.11.
- **Post-processing**: *compute (measure) terminological differences and detect terminological saturation*. These include the merging of the extracted bags of terms (for the MPCV method, Sect. 4.3); terms grouping (Sect. 4.5); the computation of terminological differences; saturation analysis; and regular accumulated noise removal (if required). The decision to use terms grouping is made while configuring the computation flow. Based on this decision, either the baseline **`THD`** (Sect. 4.5) or refined (Sect. 4.6.4) **`R-THD`** algorithm is used for computing terminological differences. Saturation analysis is done manually (greyed in Fig. 4.1) using the terminological difference reports generated by **`THD`** or **`R-THD`** (see Fig. 4.2). As a result, the accumulation of regular noise might be detected (c.f. Sect. 5.5.3). In this case, the de-noising of the bags of terms is performed with partial automation (Sect. 4.7).

[1] The specification of the used BPMN 2.0 could be retrieved from https://www.omg.org/spec/BPMN/2.0/PDF.

Dataset:	Extracted:	Score > 1:	Retained:	eps:	thd:	thdr
D1.txt:	39132:	20609:	1864:	12.0:	62.89:	100
D2.txt:	63676:	33854:	2537:	15.5:	31.50:	43.68
D3.txt:	85596:	46149:	3022:	18.0:	23.38:	32.93
D4.txt:	109076:	58400:	3399:	19.65:	18.03:	27.23
D5.txt:	135006:	71194:	3983:	23.22:	15.80:	21.43
D6.txt:	152777:	80495:	4279:	24.0:	10.06:	13.78
D7.txt:	173655:	91576:	4954:	24.0:	10.67:	13.59
D8.txt:	193998:	102449:	5020:	26.0:	9.13:	11.32
D9.txt:	215213:	113413:	5345:	28.0:	8.68:	10.33
D10.txt:	237382:	124313:	5656:	28.52:	7.56:	8.69
D11.txt:	258522:	135172:	6177:	28.52:	7.30:	8.13
D12.txt:	275178:	144390:	6625:	28.52:	6.53:	6.97
D13.txt:	291482:	152890:	6661:	30.0:	6.43:	6.81
D14.txt:	306387:	161301:	5685:	38.0:	15.46:	15.87
D15.txt:	313506:	164726:	5815:	38.0:	3.52:	3.66

Fig. 4.2 An example of a terminological difference report for the sequence of incrementally extended datasets

The workflow, pictured in Fig. 4.1 and discussed above, is an activity-level view on the process of computation for saturated terminology extraction. The programmatic details, required for its implementation in software, are presented in the structured computation flow diagram of the saturated terminology extraction pipeline in Fig. 4.3.

4.2 Preparatory Steps and Algorithms

The preparatory phase includes the steps of Document Sampling and Catalogue Generation, and Document Retrieval (Fig. 4.3).

4.2.1 Catalogue Generation

For document sampling, the method and software by Dobrovolskyi and Keberle [1] are used.

For generating the catalogue of the sampled document collection, the document repository needs to be parsed and the metadata for every document, included in the collection, appended to the collection catalogue. In addition to metadata, the information about the number of citations for every document needs to be retrieved.

A collection catalogue is a two-dimensional table with the rows representing documents and columns—the attributes of these documents. The attributes are explained in Table 4.1.

The high-level view on the algorithm for generating the catalogue for a document collection is presented in Algorithm 4.1.

Algorithmically, the task of generating the catalogue of a document collection is repository-specific. The specificity lies in the way a specific repository presents

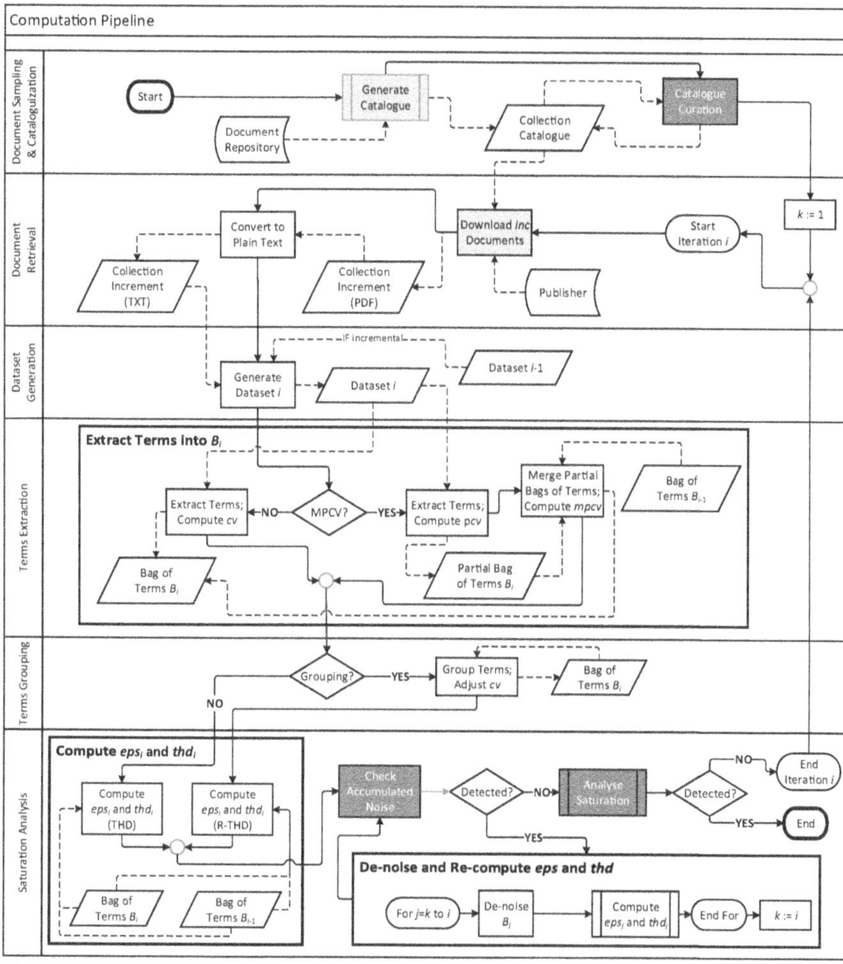

Legend: Light-greyed modules and sub-processes either require user interaction, or are performed partially manually. Dark-greyed tasks and sub-processes are manual.

Fig. 4.3 The computation flow diagram of the process of saturated terminology extraction

the documents and allows access to their metadata using their particular application programming interface (API).

Hence, the functions (**`DR-query()`**, **`parse()`**), used in the **`GCAT`** algorithm (Algorithm 4.1), are document repository API specific in their implementation. Similarly, the format of the input query string Q (the second input parameter of **`GCAT`**) is document repository API specific. In our work, two different implementations

Table 4.1 The attributes of documents in a collection catalogue

Column	Header (attribute)	Explanation
A	Publication year	The year of publication of the document
B	Type	One of the following values to be chosen based on the type of the document: UN—unknown, JA—journal article, PT—patent, CP—conference paper, BC—book chapter, BO—book, BR—book reference, DA—dataset, RE—repository
C	Venue (journal/series/conference)	The value is: (i) the name of the journal if the document is a journal article (JA); the name of the book series (e.g. LNCS) if the doc is a book chapter (BC); conference name + book series (e.g. ISWC 2020; LNCS) if the doc is a conference or workshop paper (CP)
D	Publisher	The name of the publisher (e.g. Springer-Nature, Elsevier, ACM, etc.)
E	Volume No	This is the journal or series volume number, only for JA, CP, BC, BO
F	Issue No	The number of the journal issue, only for JA
G	Pages	For JA, BC, CP, the starting-ending pages within the volume. For electronic publications, these might be paper no. and no. of pages. For manuscripts, like theses—the total no of pages
H	DOI	The digital object identifier (DOI) of the document (if available)
I	DOI link	The URL, containing the DOI and pointing to the original publication at the publisher's resource (if available)
J	MSR ID	The identifier of the document in the MS research (MSR) repository (if available)
K	Title	The title of the document, including the sub-title, if any
L	Authors	The semicolon-separated list of the authors of the document in MSR format
M	Affiliations	The semicolon-separated list of the author affiliations in MSR format
N	Complete citation	The complete citation of the document in one of the bibliographic formats, e.g. BiBTeX
O	Abstract	The abstract of the document (if available)
P, Q, R	Category	These attributes are to be used for the categorization or partitioning of the collection, which is application specific
T	– Reserved –	

(continued)

Table 4.1 (continued)

Column	Header (attribute)	Explanation
U	Citation count (GS)	The number of document citations acquired from Google Scholar (GS)
V	Citations per year (GS)	The citation frequency for document ranking/ordering, computed as: the formula is: Citation-Count-(U)/(<current year > -Publication-Year-(A) + 1)
W	Document file name	The file name for the full text file of the document that is further downloaded and stored for processing. The rule for generating the name is as follows: <Year>+"- "+<Type>+"- "+<Volume>+"("+<Issue>+")-("+<Pages>+")- "+ (substring of <DOI> after '/'; all the <DOI> string if there is no '/' in it) Year: Publication year, attribute A Type: Type, attribute B Volume: Volume No, attribute E, 1 if not available Issue: Issue No, attribute F, 1 if not available Pages: attribute G DOI: attribute H
X	MSR Entry (URL)	The URL for the document description at MSR
Y	Full text download (URL)	The URL for downloading the full text of the document, if available

have been done: (i) for the Springer-Nature Journals repository[2]; and (ii) Microsoft Research repository[3]—please see also Sect. 4.8.

Algorithm 4.1 The GCAT algorithm for generating the catalogue of a document collection

```
Input:
 DR-URL - the URL of the document repository
 Q      - the query (string) to select documents from the document
          repository
Output:
 CAT    - the XSLX file of the catalogue in the tabular form,
          containing the entries bibliographic information about
          the documents included in the collection.

1. initialize(CAT)
//  Query the DR, get the result, as the list of URL of document
//  descriptions (metadata)
2.  List := DR-query(DR-URL, Q)
3.  initialize(CAT [1..|List|])
4.  for i := 1 to |List|
      //  Acquire metadata attributes
5.    parse(List[i], CAT[i])
      //  Acquire citation count from Google Scholar
6.    CAT[i].Citation-Counts-(GS) := GS-query(CAT[i].Title,
                                              CAT[i].Authors,
                                            CAT[i].Publication-Year)
      //  Compute citation frequency
7.    CAT[i].Citations-per-Year-GS := CAT[i].Citation-Counts-GS /
                              ( year() - CAT[i].Publication-Year + 1 )
      //  Generate file name
8.    CAT[i].Document-File-Name := CAT[i].Publication-Year + "-" +
                                   CAT[i].Volume-No + "(" +
                                   CAT[i].Issue-No + ")-(" +
                                   CAT[i].Pages + ")-" +
                                   substr(CAT[i].DOI, "/")
9.  end for
10. return(CAT)
```

4.2.2 Documents Download

The step of downloading full-text documents comprises the download of the document files, in portable data format (PDF). The algorithm for this computational activity is presented in Algorithm 4.2.

[2] https://link.springer.com/journals/.

[3] https://academic.microsoft.com/.

Algorithm 4.2 The DDL algorithm for downloading a full-text document from a document repository

```
Input:
 CAT     - the collection catalogue - an XSLX file
           of the catalogue containing the entries
           of bibliographic information about the
           documents included in the collection.
Output:
 PD      - the name of the local directory to save downloaded
           PDF files.

1.   if (check-dir(PD) = .false.)
       then create-dir(PD)
       else prune(PD)
2.   for i:= 1 to |CAT|
3.     download(CAT[i].Full-text-download-URL,
                // URL to download from
                CAT[i].Document-File-Name+".PDF",
                // file name to assign
                PD)
                // directory name to save the file
4.   end for
5.   return (.true.)
```

The **DDL** algorithm is part of a higher-level batch download procedure in the computation pipeline. Hence, the implementation of the **download()** function for a particular repository has to avoid sending too many download requests without waiting until the query queue empties or shrinks to the size admissible by the repository service. The procedure of submitting bulk download requests is normally regulated by the conditions of the use of the repository, which often specifies a time interval to wait before submitting the subsequent request in the bulk.

4.3 Pre-processing Steps and Algorithms

The pre-processing phase comprises the conversion of the downloaded PDF files to plain texts, the configuration of the computation flow, and the generation of the datasets from the plain text documents.

4.3.1 Documents Conversion (PDF to Plain Text)

As terminology extractors, including those used in our work, take in only plain text files, the downloaded full-text documents need to be converted to plain texts in a way to enable the maximal possible quality of the linguistic processing pipeline of the chosen extractor software. Therefore, the following requirements need to be met by the convertor software:

- Remove hyphenations and line splits after hyphenations.
- Make every sentence start at a new line.
- Replace all ligatures in the text with corresponding letter sequences. Ligature table to be read from a separate file.
- Save the plain text output encoded as admissible by the extractor software.

As one of the extractors, used in our work, takes in ASCII[4] [2] files and the other takes in UNICODE,[5] the output encoding is passed as a parameter value to the **PDF2TXT** algorithm—see Algorithm 4.3. The algorithm performs the bulk conversion of the PDF files in the input directory and saves the TXT files in the output TXT directory with the same file names.

Algorithm 4.3 The PDF2TXT algorithm for extracting the texts from the set of PDF full text files into plain text files

```
Input:
 PD     - the name of the local input directory containing
          the PDF files to be converted to plain texts.
 TD     - the name of the local output directory for saving
          the converted TXT files.
 ENC    - the output encoding: "ASCII" or "UNICODE".
Output:
 TXT files in the TD directory containing the plain texts
 of the respective PDF files in the PD directory and having
 the same names as the PDF files.

//  Check input parameters and Initialize
1.  if (check-dir(PD) = .false.) then return (.false.)
2.  if (check-dir(TD) = .false.)
      then create-dir(TD)
      else prune(TD)
3.  if (ENC <> "ASCII" or ENC <> "UNICODE")
      then ENC := "UNICODE"
//  Process files from PD
4.  while ((FN := nextfile(PD)) <> "")
5.    PDFF := open-file(PD + FN + ".PDF", "read")
6.    TXTF := new-file(TD + FN + ".TXT", "write")
```

[4] http://ascii-table.com/ansi-codes.php.

[5] "The Unicode Standard, Version 13.0—Core Specification Appendix C" (PDF). *Unicode Consortium*. Retrieved 2020-03-11. https://unicode.org/charts/.

```
     RawTxt := extract-txt(PDFF) // An off-the-shelf
                                 // API function call
                                 // to extract the textual
                                 // part of the PDF file
                                 // in UNICODE
     //  Process the raw text
     //  move 1st sentence from RawTxt to to NextSentence
     while ((NextSentence := extract-sentence(RawTxt))<>"")
       // remove line breaks and hyphens
       NextSentence := nl-hph-remove(NextSentence)
       // replace ligatures by corresponding letter
          sequences
       NextSentence := lig-replace(NextSentence)
       // encode as ENC
       if (ENC <> "ASCII") then unicode2ascii(NextSentence)
       // write the sentence to the txt output and add
          the line break
       write(TXTF, NextSentence + NL)
     end while
     close-file(TXTF)
   end while
   return (.true.)
```

4.3.2 Configuration and Datasets Generation

The computational pipeline for saturated terminology extraction (c.f. Fig. 4.3) contains several logical forks in the computation flow. The choices of the paths in these forks and some other computational parameters are a matter of configuration. Table 4.2 presents the list of the configuration parameters, their admissible values, and the rationale for the choices. The parameters are split into two groups based on their foci on:

- Shaping out the flow of computation; and
- Grouping the documents in the sequence of plain text datasets for saturated terminology extraction.

The configuration of the computation flow is performed:

- Either using a simple GUI form with the subsequent check of the entered parameters for their conformance with the admissible values and cross-parameter relationships.
- Or by entering the proper values in the scripts invoking software modules implementing the relevant algorithms of the pipeline.

The generation of a dataset for the current iteration of the computation flow is performed using the **GDS** algorithm presented in Algorithm 4.4.

Table 4.2 Configuration parameters for the saturated terminology extraction pipeline

#	Configuration parameter	Admissible values	Rationale for the choice of values
1	*Computation flow*		
1.1	UTG—use of terms grouping	*true*, *false*	UTG = *true* is chosen if the terms grouping step is included in the pipeline. If true, then the following parameters have to be set: TSM, TST. The choice of this value invokes the execution of the **STG** algorithm (Sect. 4.6.3) before and the execution of the **R-THD** algorithm (Sect. 4.6.4) while computing terminological differences UTG = *false* is chosen if terms grouping is deemed as unnecessary. If this value is chosen, the baseline **THD** algorithm (Sect. 4.5) is executed for computing terminological differences, which uses character string equivalence for matching terms
1.2	TSM—term similarity measure	jr, jw, ja, sd	TSM = jr means that Jaro similarity measure is used to group similar and match terms. This measure is operationalized in the **M-JR** algorithm TSM = jw means the use of Jaro-Winkler similarity (**M-JW** algorithm) TSM = ja means the use of Jaccard similarity (**M-JA** algorithm) TSM = sd means the use of Sørensen-Dice similarity (**M-SD** algorithm) These algorithms are presented in Sect. 4.6.3 The selection of these measures is argued in Sect. 4.6.1
1.3	TST—term similarity threshold	A value from]0, 1[	The choice of a threshold for the chosen similarity measure is explained in Sect. 4.6.2
1.4	CVM—the method for computing *C*-Values	cv, mpcv	CVM = cv invokes the baseline method and use of incrementally extended datasets (DP = *false*). This baseline method is substantially more expensive computationally, as proven in Sect. 5.7 CVM = mpcv invokes the optimized method based on the use of the partitions of the document collection (Sect. 3.6). This optimized method employs, as an additional step before computing the terminological differences, the merging of the partial *C*-Values (Sect. 4.4.2, MPCV algorithm). This method requires that the datasets be generated as the partitions of the document collection: DP = *true*

(continued)

Table 4.2 (continued)

#	Configuration parameter	Admissible values	Rationale for the choice of values
2	*Datasets*		
2.1	DO—document ordering for forming datasets	ch, rch, rn, bd, dcf	DO = ch means that the documents are taken to dataset increments in chronological order, based on their timestamp of publication DO = rch stands for the reversed-chronological order DO = rn stands for the random choice of the documents DO = bd means that the documents are taken bi-directionally—the newest, the oldest, etc. DO = dcf stands for the use of the descending citation frequency order
2.2	INC—increment size for the sequence of datasets	A natural number	INC—specifies the number of documents that form at increment to the document sub-collection in the successive approximation process of building the terminological core collection (c.f. Sect. 3.1). These documents are taken from the collection and added to the subsequent dataset in the order specified by DP
2.3	DP—document collection partitioning	*True*, *false*	If DP = *false*, the dataset $D_{i+1} = D_i \cup INC$, where *INC* is the increment containing several (*inc*) document texts, chosen from the collection in the order, specified by DO. Hence, D_{i+1} is the incremental extension of D_i If DP = *true*, the dataset $D_{i+1} = INC$. Hence D_i and D_{i+1} are different partitions of the collection and their intersection (in documents) is empty

Algorithm 4.4 The GDS algorithm for generating the dataset for the current iteration of the computation flow

```
Input:
  CAT    - the XSLX file of the collection catalogue.
  IN     - the number of iteration
  INC    - the size of the increment (configuration parameter)
  TD     - the path to the local directory containing
           the TXT files
  DD     - the path to the local directory for storing
           the generated dataset
  DO     - the ordering of the documents selected
           for the increment (configuration parameter)
  DP     - collection partitioning (configuration parameter).
Output:
  DI     - the dataset file saved in DD.
//  Initialize and check parameters
1.  if (IN = 1) then cat-sort(CAT, DO)
2.  if (check-dir(DD) = .false.) then create-dir(DD)
3.  if ((IN-1)*INC) >= |CAT|)
      then return ("No documents for iteration" + str(IN))
// Generate dataset
4.  DI := "D" + str(IN, 4);
    DIF := open-file(DD + DI + ".TXT", "rewrite")
5.  if ((not DP) and IN > 1)
      then
        // Incremental: DI contains DI-1 plus the increment
6.      DI-1F := "D" + str(IN - 1, 4)
7.      DI-1F := open-file(DD + DI-1F + ".TXT", "read")
8.      copyfile(DI-1F, DIF)
9.  end if
// Append increment
10. for i := ((IN - 1) * INC + 1) to min(IN*INC, |CAT|)
11.   D := readfile(TD + CAT[i].Document-File-Name+".TXT")
12.   write(TXTF, D)
13. end for
14. return (DD + DI-1F + ".TXT - generated")
```

4.4 Algorithms for the Optimized Computation Pipeline

Two aspects have been identified in our work that helps substantially reduce the computation time of the statistical part of the terms extraction pipeline for terminological saturation detection. In Sect. 2.11, we argued that the use of the Aho-Corascik algorithm, for efficient matching of multiple character strings that nest the input set of strings, significantly lowers the computational expense for computing *C*-Values. Further, in Sect. 3.6, we elaborated a computationally efficient method for computing merged partial *C*-Values based on the use of the partition of a document collection. In this section, we develop the algorithms for these refinements of the baseline *C*-Value computation method that uses incrementally extended datasets.

4.4.1 Use of the Aho-Corascik Algorithm in Computing C-Values

Aho and Corascik [3] proposed to generate a trie data structure and use a Final State Machine (FSM) for efficient searching character substrings from a given set (a dictionary) in longer text strings. Their results prove the five to ten times increase of performance, compared to a conventional substring search. The approach by Aho and Corascik has been integrated into our baseline algorithm for computing *C*-Values. The result is presented in Algorithm 4.5.

Algorithm 4.5 The AC-CV algorithm for optimized computing of *C*-Values using Aho-Corascik algorithm [3] for fast multiple string matching

```
Input:
 S    - The bag of candidate term strings extracted by
        the linguistic part of the pipeline. In the bag,
        a candidate string is repeated as many times as it
        appears in the textual dataset.
Structures:
 B    - The bag of extracted term candidates. Every B[i]
        is a structure containing (term:string, cv:float),
        where: term is a term candidate string that appears
        only one time in B; cv is its C-Value to be computed.
 ST   - the table of candidate term strings and nesting
        candidate term strings built using the Aho-Corascik
        FSM. Columns: ST.term; ST.nstterm, ST.nstcv
Output:
 B    - the bag of extracted terms

//  Initialize
1.  A := ahocorascik.automaton() // used algorithm
                                 // Methods: A.add_word();
                                 // A.make_automaton();
                                 // A.iter()

// (1) Compute self-frequencies, form B
2.  sort(S)
3.  k := 0; s := ""; f := 0
4.  for i :=1 to |S|
5.   if (s <> S[i])
```

```
        // New candidate term (B[k].term) in the bag of terms.
        // It is so far assumed that s is not nested in the
        // rest of B
        // Term frequency f is stored in B[k].cv at this step
        then k += 1; append(B); f := 1; s := S[i];
             B[k].term := s; B[k].cv := f
        // The same candidate term
        else f += 1
    end for

// (2) Generate the Aho-Corasick trie and FSM for searching
//     nested terms
    for i := 1 to |B|
      A.add_word(B[i].term, (i, B[i].term, B[i].cv))
    end for
    A.make_automaton()

// (3) Build and index the table of nestings ST
    for i := 1 to |B|
      cur-term := B[i].term
      cur-cv := B[i].cv
      for (end-index, (insert-order, original-value))
      in A.iter(cur-term)
        if (original_value <> cur-term)
          then append(ST, (original-value, cur-term, cur-cv))
      end for
    end for
    IST := index(ST, ST.term+ST.nstterm)

// (4) Adjust the C-Values in B[.].cv using ST
    for i := 1 to |B|
      k := 0; cval := 0
      for ind in loc(IST, ST.term, B[i].term)
        cval -= ST[ind].nstcv; k += 1
      end for
      if (k > 0) then cval /= k
      B[i].cv := log(len(B[i].term) * (B[i].cv + cval)
    end for
    return (B)
```

4.4.2 Merging Partial C-Values

The merging of partial *C*-Values (c.f. Sect. 3.6, Eqs. 3.29 and 3.31) is required for computing $mpcv(s)$, which do not practically differ from conventionally computed *C*-Values, as proven in Theorem 3.5. This merging for the sequence of the sets of extracted terms, $B_i, i = 1, \ldots, n$, where n is the number of partial collections of a document collection, is done by iterating the **MPCV** algorithm,[6] presented as Algorithm 4.6.

Algorithm 4.6 The MPCV algorithm for merging partial *C*-Values from the two bags of terms

```
Input:
 Bi, Bi+1 - the bags of extracted terms. Each term Bi.term
            is accompanied with its Bi.pcv. Bi, Bi+1 are
            sorted in the descending order of.pcv.
Output:     The bag of terms Bi+1 merged with Bi into Bi+1.
    resort := .false.
    for k := 1 to |Bi|
      match := .false.
      for m := 1 to |Bi+1|
        if (Bi[k].term = Bi+1[m].term)
           then Bi+1[m].pcv += (Bi[k].pcv; match := .true.
        if (match <> .true.)
           then append(Bi[k], Bi+1); resort := .true.
      end for
    end for
    if (resort) then sort(Bi+1, Bi+1.pcv, desc)
    return (Bi+1)
```

4.5 Baseline Algorithm for Terminological Difference Measurement

The baseline algorithm (**THD**) for computing terminological difference *thd* between the two bags of retained significant terms T_i and T_{i+1} (Sect. 3.3) has been proposed by Tatarintseva et al. [5]. This section provides the description of this algorithm, for the sake of completeness, in a refined and extended form and different notation (Algorithm 4.7). In addition to [5], this presentation includes the computation of the individual term significance threshold (*eps*) for the bag of extracted terms B_{i+1} and, hence, the extraction of T_{i+1}. It is assumed that T_i is taken in as it has been computed in the previous iteration of the computation flow.

The baseline **THD** algorithm accumulates the *n-score* differences, between the sets of terms in T_i and T_{i+1}, in the thd value for T_{i+1}. For matching the terms, their character strings are literally compared (c.f. line 7 in Algorithm 4.7). If there was no literally matching term in T_i, the *n-score* of this new term is added to the *thd* value of T_{i+1}. Otherwise, if a match has been found, the difference in *n-score*s is added to

[6] The **MPCV** algorithm has been published by Kosa et al. in [4]. It is given here for the sake of completeness of the presented Algorithmic Suite.

the *thd* value of T_{i+1}. Finally, the relative terminological difference *thdr* is computed as *thd* divided by the sum of *n-scores* in T_{i+1}.

Absolute (*thd*) and relative (*thdr*) terminological differences are further used for finding out if the term-set in T_{i+1} differs from the term-set in T_i for more than the individual term significance threshold eps_{i+1}. If not, it indicates that adding an increment of documents to D_i, resulting in D_{i+1}, did not bring any noticeable amount of new terminology. It implies from this observation that the sub-collection D_{i+1} of the overall collection may have become terminologically saturated. However, as specified in our terminological saturation condition (ii) in Eq. (3.4), it is required that in all the subsequent iterations thd_{i+1} remains below eps_{i+1}. Therefore, to become more certain about the saturation, it is recommended that more subsequent pairs of T_i and T_{i+1} are evaluated. If the second condition (3.4) is steadily fulfilled, then the process of looking for a minimal saturated sub-collection could be stopped.

Algorithm 4.7 The baseline THD algorithm, adapted from Tatarintseva et al. [5]

```
Input:
 Ti    - the bag of retained significant terms of the previous
         iteration
 Bi+1  - the bag of extracted terms of the current iteration.
 In these bags of terms, every term Ti.term or Bi+1.term is
 accompanied by its C-Value: Ti.cv or Bi+1.cv. Ti and Bi+1
 are sorted in the descending order of C-Values.
Output: Ti+1, eps, thd, thdr

//    Compute eps for Bi+1, and Ti+1
      sum := 0
      for k := 1 to |Bi+1|
        sum +=  Bi+1[k].cv
      end for
      eps := 0.0; count :=0
      for i := 1 to |Bi+1|
        eps += Bi+1[k].cv
        if (eps > (sum / 2.0))
          then
            eps := Bi+1[k].cv
            while eps = Bi+1[count+1].cv
              count += 1; append(Bi+1[count], Ti+1)
```

```
            end while
            break
          else
            count += 1; append(Bi+1[count], Ti+1)
      end for

//    Compute thd and thdr
      sum := 0
      thd := 0
      for k := 1, |Ti+1|
        sum += Ti+1[k].cv
        found : = .false.
        for m := 1, |Ti|
          if (Ti+1[k].term = Ti[m].term)
            then
              thd += |Ti+1[k].cv - Ti[m].cv|
              found := .true.
        end for
        if (found <> .true.) then thd += Ti+1[k].cv
      end for
      thdr := thd  / sum

      return (eps, Ti+1, thd, thdr)
```

4.6 Algorithms for Terms Grouping

One research question (**RQ8**, c.f. Sect. 2.13) has not been given an answer in the formal framework developed in Chap. 3. The question was if grouping similar terms might facilitate faster and less volatile convergence to terminological saturation. In this section, we formulate the research hypothesis to answer this question as follows.

Hypothesis H4.1 Grouping similar terms leads to faster terminological saturation and more compact sets of retained significant terms.

Let a bag of extracted terms contain the subsets of candidate strings that are similar in their intended meaning. Then each of such subsets could be grouped in one candidate string and its *C*-Value adjusted to reflect the consolidated significance of the subset. For that: (i) the relevant measures of term similarity have to be chosen; (ii) the thresholds to distinguish between similar and non-similar candidates have to be rationally proposed; and (iii) the algorithms for similarity measurement, terms grouping, and refined terminological difference measurement have to be developed. The validity and utility of this approach have to be further experimentally evaluated.

Based on the outline of the approach in **H4.1**, we elaborate it in this section. Experimental evaluation is done in Sect. 5.6.

For elaborating the approach,[7] we first select four string similarity measures from those reviewed in Sect. 2.10. We then deliberate on the possible cases of how partial

[7] The material of this section is based on our publication [6].

string similarity might be regarded as corresponding term (meaning) similarity. Based on this deliberation and a manually collected test set of term pairs, we devise the similarity thresholds for all four selected string similarity measures. Using the formal representations of these measures and thresholds, the algorithms are developed for computing:

- The measures: **M-JR**, **M-JW**, **M-JA**, **M-SD**
- Similar term groupings: **STG**.

Finally, the refined algorithm, **R-THD**, is developed for measuring terminological differences. In difference to **THD** (Algorithm 4.7), **R-THD** uses string similarity measures and corresponding term similarity thresholds.

4.6.1 Choice of String Similarity Measures

From the variety of SSM, mentioned in Sect. 2.10, due to the specifics of our task of the approximate comparison of short strings containing a few words, the following SSM were filtered out: (i) those that required long strings or sets of strings of a considerably big size; (ii) those that were computationally hard. The representation of all kinds of string similarity measures in our short list was also intended to be kept, if appropriate. As a result, the following list of measures to be considered for further use was formed:

- Character-based measures: Levenshtein distance [7], Hamming distance [8], Jaro similarity [9], and Jaro-Winkler similarity [10].
- Token-based measures: Jaccard similarity for uni-gram comparison [11], cosine similarity for uni-gram comparison [12], and Sørensen-Dice coefficient for bi-gram comparison [13, 14].

Among those, Levenshtein and Hamming distances appeared to be the least appropriate in our context due to their specifics. Levenshtein returned an integer number of required edits, while the rest of the measures returned normalized reals. Hence, it was not clear if normalizing Levenshtein would have made the result comparable to the other measures in a way to use the same term similarity threshold. Hamming measure was applicable only to the strings of equal lengths. Adding spaces to a shorter string, however, may have lowered the precision of measurement. Cosine similarity was based on the same principle as Jaccard, but was more computationally expensive due to the presence of the square root in the denominator. Therefore, we finally chose to use Jaro, Jaro-Winkler, Jaccard, and Sørensen-Dice for implementation and evaluation in our work. Further, it is briefly explained how the selected measures are computed.

Jaro similarity sim_j between two strings S_1 and S_2 is computed as the minimal number of one character transforms to be done to the first string for getting the second string in the compared pair:

$$sim_j = \begin{cases} 0 & \text{if } m = 0 \\ \frac{1}{3} * \left(\frac{m}{|S_1|} + \frac{m}{|S_2|} + \frac{m-t}{m} \right) & \text{otherwise} \end{cases}, \quad (4.1)$$

where: $|S_1|$, $|S_2|$ are the lengths of the compared strings; m is the number of the matching characters; and t is the half of the number of transposed characters. The characters are matching if they are the same and their distance from the beginning of the string differs by no more than $max(|S_1|, |S_2|)/2 - 1$. The number of transposed characters is the number of matching but having different sequence order symbols.

Jaro-Winkler similarity measure sim_{j-w} refines Jaro similarity measure by using a prefix scale value p, which assigns better ratings to the strings that match from their beginnings for a prefix length l. Hence, for the two strings S_1 and S_2 it is computed as:

$$sim_{j-w} = sim_j + l*p*(1 - sim_j), \tag{4.2}$$

where: l is the length of a common prefix (up to a maximum of 4 characters); p is a constant scaling factor meaning how much the similarity value is adjusted upwards for having common prefixes (up to 0.25, otherwise the measure can become larger than 1; Winkler [10] suggests that $p = 0.1$).

Sometimes Winkler's prefix bonus $l * p * (1 - \text{sim}_j)$ is given only to the pairs having Jaro similarity value higher than a particular threshold. This threshold is suggested to be equal to 0.7 [10].

Jaccard similarity index sim_{ja} is a similarity measure for finite sets, characters in our case. It is computed, for the two strings S_1 and S_2, as the ratio between the cardinalities of the intersection and union of the character sets in S_1 and S_2:

$$sim_{ja} = (|S_1| \cap |S_2|)/(|S_1| \cup |S_2|). \tag{4.3}$$

Finally, the Sørensen-Dice coefficient is computed by counting identical character *bi*-grams in S_1 and S_2 and relating these to the overall number of *bi*-grams:

$$sim_{sd} = 2n_{\equiv}/(n_{S_1} + n_{S_2}), \tag{4.4}$$

where: $n_{\equiv}$ is the number of *bi*-grams found in S_1 and also in S_2; n_{S_1}, n_{S_1} are the numbers of all *bi*-grams in S_1 and S_2 respectively.

4.6.2 Term Similarity Cases and Thresholds

For the proper use of the implemented SSM functions in the context of terms comparison and grouping, it is necessary to determine what would be a reasonable threshold to distinguish between (semantically) similar and different terms. For finding that out, the following cases in string comparison need to be taken into account.

Full Positives (FP). In this case, evaluated character strings are fully the same, which clearly gives similar (the same) terms.

Full Negatives (FN). In this case, evaluated character strings are very different and the terms in these strings carry different semantics. This is also a clear situation, which is characterized by the low values of similarity measures.

Partial Positives (PP). In this case, evaluated character strings are partially the same and the terms in these strings carry the same or similar semantics. The terms in such strings are similar, though it may not be fully clear. The following are different categories of terms that fall into this case:

The words in the terms have different endings (e.g. plural/singular forms).
Different delimiters are used (e.g. "-", or "–", or " - ").
A symbol is missing, erroneously added, or misspelt (a typo).
One term is the substring of the other (e.g. subsuming the second).
One of the strings contains unnecessary extra characters (e.g. two or three spaces instead of one, or noise).

Partial Negatives (PN). In this case, evaluated character strings are partially the same but the terms in these strings carry different semantics. The terms in such strings are different, though it may not be fully clear. The following are the categories that fall into this case: the terms in the compared strings differ by a very few characters, but have substantially different meanings (e.g. "deprecate" versus "depreciate"); the compared multi-word terms have common word(s) but fully differ in their meanings (e.g. "affect them" versus "effect them"). These PN are the hardest case to be detected.

The test set of term pairs falling into the cases and categories described above has been manually developed.[8] For each pair of terms in this test set, all four selected string similarity measures have been computed. The fragment is presented in Table 4.3.

The average values of all four chosen similarity measures for each category have been computed using all the test set term pairs falling into this category. These values are presented in Table 4.4.

Term similarity thresholds have to be chosen such that full and partial negatives are regarded as not similar, but full and partial positives are regarded as similar. Hence, for the case of partial positives, the thresholds have to be chosen as minimal of all the case categories, and for the partial negatives—as the maximal of all the case categories. The values of case thresholds are given in Table 4.5.

These thresholds are further used as the margins for the corresponding threshold intervals in our experiments. These intervals have been evenly split by the four threshold points, as presented in Table 4.5. The requirements for partial positives and negatives, unfortunately, contradict each other. For example, if a threshold is chosen to filter out partial negatives, also some of the partial positives will be filtered out. Therefore, subsuming that partial negatives are rare, it has been decided to use the thresholds for partial positives.

[8] The test set and computed term similarity values are publicly accessible at https://github.com/OntoElect/Data/blob/master/STG/Test-Set.xls.

Table 4.3 Similarity measures for different test cases [6]

Case	Category	Terms pair	Sørensen-Dice	Jaccard	Jaro	Jaro-Winkler
Different (FN)		whirled ǀ world	0.000	0.500	0.790	0.811
		traces ǀ creta	0.000	0.833	0.588	0.588
		time domain ǀ ontology lifecycle	0.000	0.428	0.445	0.445
Same (FP)		identical strings ǀ identical strings	1.000	1.000	1.000	1.000
Similar Semantics (PP)	Extra characters	*system?problems ǀ system problems	0.814	0.769	0.936	0.936
		sad data mining ǀ sqr data mining	0.769	0.818	0.859	0.873
	Common parts (words)	marcov chain monte carlo methods ǀ monte carlo methods	0.782	0.766	0.629	0.666
		data mining algorithm ǀ data mining	0.642	0.666	0.842	0.904
		cation error ǀ error	0.533	0.333	0.427	0.427
	Typos	fraud detection ǀ froud dit ection	0.714	0.916	0.859	0.887
		monte carlo ǀ monte ??rlo	0.700	0.727	0.878	0.927
		data mining ǀ data minin	0.941	0.875	0.969	0.981
	Different delimiters	computer science ǀ computerscience	0.896	0.916	0.979	0.987
		serial episodes ǀ serial&&episodes	0.827	0.818	0.936	0.961
		data cube ǀ data_cube	0.750	0.777	0.925	0.955
	Different endings	network structure ǀ network structures	0.969	1.000	0.981	0.988
		time complexity ǀ time complexities	0.896	0.833	0.981	0.951
		value ǀ values	0.888	0.833	0.918	0.951
Different Semantics (PN)	Common parts (words)	database ǀ military base	0.400	0.500	0.410	0.410
		brainstorm ǀ stormy weather	0.363	0.428	0.509	0.509
		iron clad ǀ iron maiden	0.444	0.636	0.804	0.882

(continued)

Table 4.3 (continued)

Case	Category	Terms pair	Sørensen-Dice	Jaccard	Jaro	Jaro-Winkler
		jellyfish | fish tank	0.352	0.307	0.614	0.614
		four delegates | delegated authority	0.451	0.666	0.557	0.557
		string theory | string format	0.583	0.571	0.812	0.887
	Very few character differences	deprecate against | depreciate against	0.909	1.000	0.903	0.941
		alternately move | alternatively move	0.933	0.916	0.900	0.940
		affect them | effect them	0.900	0.758	0.906	0.906

Table 4.4 Average values of string similarity measures for different categories of term pairs from the test set [6]

Case/category	Items in test set	Sørensen-Dice	Jaccard	Jaro	Jaro-Winkler
Different strings (FN)	6	0.03	0.45	0.55	0.55
Identical strings (FP)	3	1.00	1.00	1.00	1.00
Similar semantics (PP)	32	0.71	0.72	0.63	0.70
• Unnecessary (extra) characters	7	0.8401	0.8820	0.8714	0.8784
• Common parts (words)	6	0.7122	0.7280	0.6375	0.7043
• Typos	6	0.7797	0.8637	0.8863	0.9220
• Different delimiters	6	0.7860	0.8473	0.9125	0.9442
• Different endings	7	0.8911	0.9135	0.9410	0.9590
Different semantics (PN)	18	0.89	0.89	0.89	0.91
• Common parts (words)	11	0.4336	0.5221	0.6161	0.6408
• Very few character differences	7	0.8826	0.8845	0.8914	0.9059
Total	59				

Table 4.5 Term similarity thresholds chosen for experimental evaluation [6]

Method	Term similarity thresholds			
	Min	Ave-1	Ave-2	Max
Sørensen-Dice	0.71	0.76	0.83	0.89
Jaccard	0.72	0.77	0.83	0.89
Jaro	0.63	0.72	0.80	0.89
Jaro-Winkler	0.70	0.77	0.84	0.91

4.6.3 Terms Grouping and Similarity Measurement

Our task was to refine the baseline **THD** algorithm (Sect. 4.4) in a way to allow finding not exactly the same but sufficiently similar terms by applying string similarity measures (Sect. 4.6.1) with appropriate thresholds, as explained in the previous Sect. 4.6.2. For that, the preliminary similar term grouping step was introduced to avoid duplicate similarity detection. For each of the compared bags of terms T_i and T_{i+1} the similar term grouping (**STG**) algorithm is applied at this preliminary step—see Algorithm 4.8. The algorithms for computing selected string similarity measures are presented as Algorithms 4.9–4.12.

Algorithm 4.8 Similar Term Grouping (STG) algorithm [6]

```
Input:
 T  - a bag of terms. Each term T.term is accompanied
      with its C-Value T.cv. T is sorted in the descending
      order of T.cv.
 M  - the name of the string similarity measure function
      to compare terms
 th - the value of the term similarity threshold from
      within [0,1]
Output: T with grouped similar terms
    sum := 0
    for k = 1 to |T|
      term := T[k].term
      cv := T[k].cv
      count := 1
      for m = k+1 to |T|
        if M(term, T[m].term, th)
          then cv += T[m].cv; count += 1; remove(T[m])
      end for
      T[k].cv := cv / count
    end for
    return (T)
```

M-JR (Algorithm 4.9) computes Jaro similarity for two input character strings, S_1 and S_2 based on Eq. (4.1). The idea to use bit maps to mark the positions of the matching characters is adopted from the similar algorithm published on RosettaCode.[9] In difference to the RosettaCode algorithm, **M-JR**: (i) makes difference between shorter and longer incoming character strings; and (ii) returns not the value of Jaro similarity but the result of its comparison with the similarity threshold (th): `.false.` if the similarity value is below the threshold or `.true.` otherwise.

Algorithm 4.9 The algorithm (M-JR) for computing Jaro string similarity measure

[9] https://rosettacode.org/wiki/Jaro_distance.

```
Input:
 S1, S2    - character strings to be compared
 th        - the value of the similarity threshold,
             within [0,1]
Structures:
 BSl,BSs - Boolean arrays with dimensions [1..|l|] and
           [1..|s|], where l is the longest and s is the
           shortest of S1, S2, indicating the positions
           of matches in S1 and S2 by .true.
Output: .true., if computed JaroSim ≥ th, .false. otherwise

//  Initialize and check inputs
if (|S1| = 0 or |S2|=0) then return (.false.)
if (th < 0.0 or th > 1.0) then return (.false.)
md := ⌊|l|/2⌋ − 1    // matching distance
cm := 0              // counter for matches
ct := 0              // counter for transpositions
if (|S1| ≥ |S2|) then (l:=S1; s:=S2) else (l:=S2; s:=S1)
BSl[1..|l|] := BSs[1..|s|] := .false.

//  Count matches
for i := 1 to |s|
  for k := max(1,i-md) to min(i+md, |l|)
    if BSl[k] then iterate
    if s[i] <> l[k] then iterate
    BSl[k] := BSs[i] := .true.; cm += 1; break
  end for
end for
if (cm = 0) then return (.false.)
//  Count transpositions
k := 1;
for i := 1 to |s|
  if (BSs[i] = .false.) then iterate
  while (BSl[k] = .false.) k += 1
  if (s[i] <> l[k]) then ct += 1
end for
ct := ct / 2

//  Compute Jaro similarity
JaroSim := (cm / |s| + cm / |l| + (cm - ct) / cm)/3
if (JaroSim ≥ th)
  then return (.true.)
  else return (.false.)
```

M-JW (Algorithm 4.10) computes Jaro-Winkler similarity for two input character strings, S_1 and S_2, based on Eq. (4.2). This includes the computation of Jaro similarity and its positive adjustment based on the length of the common prefix in S_1 and S_2. Jaro similarity is computed identically to **M-JR** presented in Algorithm 4.9.

Algorithm 4.10 The algorithm (M-JW) for computing Jaro-Winkler string similarity measure

```
Input:
 S1, S2  - character strings to be compared
 th      - the value of the similarity threshold, within [0,1]
 lcp      - the length [1 … 4] of the common prefix
            to be checked in S1, S2
 sf       - the value of the scaling factor (p) in the range
            of [0.1 … 0.25]
Structures:
 BSl,BSs - Boolean arrays with dimensions [1..|l|]
            and [1..|s|], where l is the longest and s is
            the shortest of S1, S2, indicating the positions
            of matches in S1, S2 by .true.
Output:     .true., if computed JWSim ≥ th, false otherwise

// Initialize and check inputs
if (|S1| = 0 or |S2|=0) then return (.false.)
if (th < 0.0 or th > 1.0) then return (.false.)
md := ⌊|l|/2⌋ - 1     // matching distance
cm := 0               // counter for matches
ct := 0               // counter for transpositions
if (|S1| ≥ |S2|) then (l:=S1; s:=S2) else (l:=S2; s:=S1)
BSl[1..|l|] := BSs[1..|s|] := .false.
if (lcp <= 0 or lcp > 4) then lcp := 4
if (sf < 0.1 or sf > 0.25) then sf := 0.1
// Count matches
for i := 1 to |s|
  for k := max(1,i-md) to min(i+md, |l|)
    if BSl[k] then iterate
    if s[i] <> l[k] then iterate
    BSl[k] := BSs[i] := .true.; cm += 1; break
  end for
end for
if (cm = 0) then return (.false.)
// Count transpositions
k := 1;
for i := 1 to |s|
  if (BSs[i] = .false.) then iterate
  while (BSl[k] = .false.) k += 1
  if (s[i] <> l[k]) then ct += 1
end for
ct := ct / 2

// Compute Jaro similarity
JaroSim := (cm / |s| + cm / |l| + (cm - ct) / cm)/3

// Compute Winkler adjustment
// Compute the length of the common prefix
clcp := 0
for i := 1 to min(lcp,|s|)
  if s[i] <> l[i] then break
end for

// Compute JWSim
JWSim := JaroSim + clcp * cp * (1 - JaroSim)
if (JWSim ≥ th)
  then return (.true.)
  else return (.false.)
```

M-JA (Algorithm 4.11) computes Jaccard similarity for two input character strings, S_1 and S_2, based on Eq. (4.3).

Algorithm 4.11 The algorithm (M-JA) for computing Jaccard string similarity measure

```
Input:
S1, S2   - character strings to be compared
 th      - the value of the similarity threshold, within [0,1]
Output:    .true., if computed JaccardSim ≥ th,
           .false. otherwise

//  Initialize and check inputs
1.  if (|S1| = 0 or |S2|=0) then return .false.
2.  if (th < 0.0 or th > 1.0) then return .false.
3.  if (|S1| ≥ |S2|) then (l:=S1; s:=S2) else (l:=S2; s:=S1)

//  Compute Intersection size
4.  is := 0
5.  for i := 1 to |l|
6.    k := 1
7.    while k <= |s|
8.      if (l[i] = s[k]) then is += 1; remove(s[k], s); break
                         else k += 1; iterate
9.    end while
10. end for

//  Compute Jaccard similarity
11. JaccardSim := is / (|S1| + |S2|)
12. if (JaccardSim ≥ th)
      then return (.true.)
      else return (.false.)
```

M-SD (Algorithm 4.12) computes Sørensen-Dice similarity for two input character strings, S_1 and S_2, based on Eq. (4.4). This includes the generation of the *bi*-gram sets of S_1 and S_2 and the computation of Jaccard similarity of these *bi*-gram sets. Jaccard similarity is computed identically to **M-JA** presented in Algorithm 4.11.

Algorithm 4.12 The algorithm (M-SD) for computing Sørensen-Dice string similarity measure

```
Input:
 S1, S2  - character strings to be compared
 th      - the value of the similarity threshold, within [0,1]
Structures:
 bl, bs - bi-gram (2-character) substring arrays
          with dimensions [1..|l|] and [1..|s|], where l
          is the longest and s is the shortest of S1, S2
Output:   .true., if computed SDSim ≥ th, .false. otherwise

//  Initialize and check inputs
if (|S1| = 0 or |S2|=0) then return .false.
if (th < 0.0 or th > 1.0) then return .false.
if (|S1| ≥ |S2|) then (l:=S1; s:=S2) else (l:=S2; s:=S1)

//  Generate bi-gram arrays
for i := 1 to |l| - 1     // bi-grams of l
  bl[i] := l[i] + l[i+1]
end for
for i := 1 to |s| - 1     // bi-grams of s
  bs[i] := s[i] + s[i+1]
end for

//  Compute Intersection size
is := 0
for i := 1 to |bl|
  k := 1
  while k <= |bs|
    if (bl[i] = bs[k])
      then is += 1; remove(bs[k], bs); break
      else k += 1; iterate
  end while
end for

//  Compute Sørensen-Dice similarity
SDSim := is / (|bl| + |bs|)
if (SDSim ≥ th)
  then return (.true.)
  else return (.false.)
```

4.6.4 *Refined Algorithm for Terminological Difference Measurement*

After term grouping is accomplished for both bags of retained significant terms T_i and T_{i+1}, the refined **THD** algorithm (**R-THD**, Algorithm 4.13) is performed to compute the terminological difference between these bags of terms.

Algorithm 4.13 The refinement (R-THD) of the baseline THD algorithm (Sect. 4.5)

```
Input:
  T_i,     - the bag of retained significant terms
             of the previous iteration.
  B_{i+1}  - the bag of extracted terms of the current iteration.
             In these bags of terms, every term T_i.term or
             B_{i+1}.term is accompanied by its C-Value: T_i.cv or
             B_{i+1}.cv. T_i and B_{i+1} are sorted in the descending
             order of C-Values.
  M        - the method (M-JR, M-JW, M-JA, or M-SD) for matching
             similar terms
  th       - similarity threshold
Output: T_{i+1}, eps, thd(T_{i+1}, T_i), thdr(T_{i+1}, T_i)

//  Compute eps for B_{i+1}, and T_{i+1}
    sum := 0
    for k := 1 to |B_{i+1}|
      sum +=  B_{i+1}[k].cv
    end for
    eps := 0.0; count :=0
    for i := 1 to |B_{i+1}|
      eps += B_{i+1}[k].cv
      if (eps > (sum / 2.0))
        then
          eps := B_{i+1}[k].cv
          while eps = B_{i+1}[count+1].cv
            count += 1; append(B_{i+1}[count], T_{i+1})
          end while
          break
        else
          count += 1; append(B_{i+1}[count], T_{i+1})
    end for

//  Compute thd and thdr
    sum := 0
    thd := 0
    for k := 1 to |T_{i+1}|
      sum += T_{i+1}[k].cv
      found : = .false.
      for m := 1 to |T_i|
        if (M(T_{i+1}[k].term, T_i[m].term, th))
          then
            thd += | T_{i+1}[k].cv - T_i[m].cv|
            found := .true.
      end for
      if (found = .false.) then thd += T_{i+1}[k].cv
    end for
    thdr := thd  / sum
    return (eps, T_{i+1}, thd, thdr)
```

In difference to **THD** (Algorithm 4.7), **R-THD** has two more input parameters: M and th. It uses M—the method (**M-JR**, **M-JW**, **M-JA**, or **M-SD**) for matching similar terms (line 24). The threshold (*th*) is passed to `M` to distinguish between similar and not similar strings.

4.7 Algorithm for Accumulated Regular Noise Removal

Due to the incremental nature of the datasets used in our computation pipeline, it is probable that some term candidate strings, that are not real terms, will have very high frequencies of appearance in the processed datasets. Starting from some iteration (i) in the process of computation, these strings may occupy the upper part of the bag of retained significant terms T_i and, thus, supersede valid terms with lower C-Values. In our work, the situation is termed as the accumulation of regular noise (ARN). Clearly, ARN should be avoided as it substantially reduces the quality of term extraction and saturation measurement, even for moderately noisy document collections. In the experimental part of our work (Chap. 5), we identified the indicators of the presence of ARN. These are: (i) a substantial decline in the number of retained significant terms; and (ii) a substantial increase of individual term significance threshold. If these indicators are observed, the extracted T_i has to be examined and the set of the strings, that are the ARN, collected and removed from all the B_i. While the examination of T_i and identification of the set of the ARN strings have to be done manually, the removal of this noise is automated by applying the **ARNR** (Algorithm 4.14).

Algorithm 4.14 The ARNR algorithm for removing ARN strings from a bag of terms

```
Input:
 ARN - A bag of accumulated regular noise character strings
 T   - A bag of terms. Each term T[i].term is accompanied
       by its T[i].cv.
Output: T without ARN
for i := 1 to |ARN|
   for j := 1 to |T|
      if ARN[i] = T[j].term
         then remove(T[j]); break
   end for
end for
return (T)
```

4.8 Implementation in Software

The algorithms of the suite, presented in Sects. 4.2–4.7, have been implemented in the software suite for computing terminological saturation. The modules of this software suite implement all the automated and support partially automated tasks in the workflow (Fig. 4.1). The implemented functions correspond to the computation flow (Fig. 4.3). The software modules of the suite are made publicly available for academic research purposes.[10] The description of the software suite and the links to the modules are provided in Table 4.6.

4.9 Summary

This chapter focused on meeting our research objective **RO2** (c.f. Sect. 2.13). We developed an algorithmic suite for implementing our formal framework for terminological saturation detection and measurement (Chap. 3) as a computational method. In particular, the algorithms developed in this chapter and implemented in the software suite (Sect. 4.8) are used in the experimental evaluation of the developed computational method in Chap. 5. Furthermore, the software suite has been made publicly available for research purposes in order to facilitate the reproducibility of our results.

The algorithmic suite has been developed for automating or partially automating the elaborated workflow for terminological saturation measurement (c.f. Sect. 4.1, Fig. 4.1) at the task level of detail. The tasks were grouped in phases. For the purpose of the development of the algorithmic suite, the workflow has been further detailed in the computation flow (c.f. Sect. 4.1, Fig. 4.3), revealing the modules (functions), data flow and transformation, iterations, and termination criteria. The algorithms for all the modules, that require the use of software, have been elaborated in Sects. 4.2–4.7. Section 4.8 reported about the implementation of the algorithmic suite in the software. By developing and implementing the algorithmic suite, we operationalized our instruments required to accomplish the full answers to the research questions **RQ1**, **4**, **5**, and **7** (c.f. Sect. 2.13.2, Table 2.5).

Yet furthermore, in Sect. 4.6, the formal and algorithmic approach to answer our research question **RQ8** about the utility of terms grouping for terminological saturation measurement was elaborated. Based on the research hypothesis **H4.1** put forward in Sect. 4.6, the four relevant string similarity measures were selected, term similarity thresholds for the selected string similarity measures were devised, and necessary algorithms for similarity measurement, terms grouping, and terminological difference measurement were developed.

The elaboration of all the mentioned components of the algorithmic suite allowed us to meet our research objective **RO2**.

---

[10] The software is licensed under the Apache License, Version 2.0 http://www.apache.org/licenses/LICENSE-2.0.

Table 4.6 The modules of the instrumental software suite

Functionality	Language	Access	Constraints
Phase[a]*: preparatory; function*[b]*: generate catalogue; algorithm:* **GCAT** (Sect. 4.2.1)			
Takes in the URL of a Springer journal page. Parses the pages of the journal issues and extracts the required metadata into the catalogue. Takes citation counts from Google Scholar[c]	PHP	https://github.com/bwtgroup/SSRTDC-Springer-article-parser	Only for Springer Link[d] journals repository
Picks the metadata from the PDF files of the processed papers. Takes citation counts from Google Scholar	Java	https://github.com/OntoElect/Code/tree/master/CatGen-PDF	The collection of PDF documents needs to be pre-downloaded to a local folder
Takes in paper identifiers mined by the snowball literature sampling pipeline. Parses the Microsoft Research (MSR)[e] paper pages for required metadata. Takes citation counts from Google Scholar	Python 3	https://github.com/gendobr/snowball/blob/master/scripts/009_extend_items.py	Only for MSR paper repository
Phase: preparatory; function: download documents; algorithm: **DDL** (Sect. 4.2.2)			
Does batch downloads of the papers listed in the catalogue. Checks if full-text URL is provided by MSR. If not, checks Google Scholar for the URL. If not, reports failure to download	Python 3	https://github.com/gendobr/snowball/blob/master/scripts/011_download_pdfs.py	The frequency of download requests may be constrained by particular full-text repositories
Phase: Pre-processing; Function: Convert PDF to plain text; Algorithm: **PDF2TXT** (Sect. 4.3.1)			
Takes in the directory with PDF files. Outputs the directory of TXT files (UNICODE)	Python 3	https://github.com/gendobr/snowball/blob/master/scripts/013_ate_pdf2txt.py	
Takes in the directory with TXT files. Applies hyphenation and line split removal, ligature substitution. Saves texts in the required encoding	Python 3	https://github.com/gendobr/snowball/blob/master/scripts/014_ate_clear_txt.py	
Phase: pre-processing; function: generate datasets; algorithm: **GDS** (Sect. 4.3.2)			

(continued)

Table 4.6 (continued)

Functionality	Language	Access	Constraints
Takes in the group of dataset configuration parameter values (Table 4.2); (ii) the directory with TXT files; and (iii) the collection catalogue. Generates the datasets out of TXT files according to the configuration parameter values	Python 3	https://github.com/gendobr/snowball/blob/master/scripts/015_ate_generate_datasets.py	
Phase: terms extraction; function: extract terms (baseline); adopted software: UPM Term Extractor			
Takes in the directory containing the datasets. Returns the directory containing the bags of extracted terms. Uses the baseline *C*-Value method (Sect. 2.11) for the statistical part	Java	https://github.com/ontologylearning-oeg/epnoi-legacy	Processes only English texts. Dataset size to be not more than 15 Mb
Phase: terms extraction; function: extract terms (optimized); algorithm: **`AC-CV`** (Sect. 4.4.1)—*used in the statistical part of the software for optimizing the computation of C-Values*			
Takes in the directory containing the datasets. Returns the directory containing the bags of extracted terms. Uses the Aho-Corascik (Aho and Corascik 1975) optimization of *C*-Value method (Sect. 2.11) for the statistical part	Python 3	https://github.com/gendobr/snowball/blob/master/scripts/lib/ate.py	Processes only English texts
Phase: post-processing; function: merge bags of terms (optimized); algorithm: **`MPCV`** (Sect. 4.4.2)			
Takes in the two consecutive bags of terms B_i and B_{i+1} containing partial *C*-Values. Outputs B_{i+1} merged with B_i as B_{i+1}. Output contains merged partial *C*-Values	Python 3	https://github.com/gendobr/snowball/blob/master/scripts/016_ate_merge_terms_partial.py	
Phase: post-processing; function: group terms; algorithms: **`M-JR`**, **`M-JW`**, **`M-JA`**, **`M-SD`**, **`STG`** (Sect. 4.6.3)			
Takes in a bag of terms. Outputs the bag of terms with grouped similar terms. Term similarity is computed using one of the four selected measures (Sect. 4.6.1) and the corresponding similarity threshold (Sect. 4.6.2)	Python 3	https://github.com/OntoElect/Code/tree/master/STG/core	

(continued)

Table 4.6 (continued)

Functionality	Language	Access	Constraints
Phase: post-processing; function: compute terminological differences (baseline); algorithm: **THD** (Sect. 4.5)			
Takes in a bag of retained significant terms T_i and a bag of extracted terms of the next iteration B_{i+1}. Returns: eps_{i+1}, T_{i+1}, thd(T_i, T_{i+1}), and thdr(T_i, T_{i+1}). Also returns the numbers of extracted and retained terms	Python 3	https://github.com/gendobr/snowball/blob/master/scripts/lib/thd.py	
Phase: post-processing; function: compute terminological difference (refined); algorithms: **M-JR**, **M-JW**, **M-JA**, **M-SD** (Sect. 4.6.3), **R-THD** (Sect. 4.6.4)			
Takes in a bag of retained significant terms T_i, a bag of extracted terms of the next iteration B_{i+1}, the name of the method M for computing similarity of two terms, and the term similarity threshold value *th*. Returns: eps_{i+1}, T_{i+1}, $thd(T_i, T_{i+1})$, and $thdr(T_i, T_{i+1})$. Also returns the numbers of extracted and retained terms	Python 3	https://github.com/OntoElect/Code/tree/master/STG	
Phase: post-processing; function: de-noise bags of terms; algorithm: **ARNR** (Sect. 4.7)			
Takes in the bag of stop terms (compiled manually via the examination of a bag of terms with accumulated regular noise), the directory with the bags of terms to be de-noised, returns the de-noised bags of terms in the same directory	Python 3	https://github.com/gendobr/snowball/blob/master/scripts/017_ate_clear_terms.py	

[a] A phase in the table corresponds to a sub-process in the workflow diagram, Sect. 4.1, Fig. 4.1.
[b] A function corresponds to a module in the computation flow diagram, Sect. 4.1, Fig. 4.3.
[c] https://scholar.google.com/
[d] https://link.springer.com/journals/
[e] https://academic.microsoft.com/

References

1. Dobrovolskyi H, Keberle N (2018) Collecting seminal scientific abstracts with topic modelling, snowball sampling and citation analysis. In: ICTERI 2018: 14th international conference on ICT in education, research, and industrial applications. Volume I: Main conference, Kyiv, Ukraine, 2018, CEUR-WS, vol 2105, pp 179–192
2. American National Standard for Information Systems—Coded Character Sets—7-Bit American National Standard Code for Information Interchange (7-Bit ASCII), ANSI X3.4-1986. American National Standards Institute (ANSI). 26 March 1986
3. Aho AV, Corasick MJ (1975) Efficient string matching: an aid to bibliographic search. Commun ACM 18(6):333–340
4. Kosa V, Chaves-Fraga D, Dobrovolskiy H, Ermolayev V (2020) Optimized term extraction method based on computing merged partial *C*-values. In: Ermolayev V et al (eds) Revised selected papers of ICTERI 2019. Communications in computer and information science, vol 1175. Springer, Cham, pp 24–49. https://doi.org/10.1007/978-3-030-39459-2_2
5. Tatarintseva O, Ermolayev V, Keller B, Matzke W-E (2013) Quantifying ontology fitness in OntoElect using saturation- and vote-based metrics. In: Ermolayev V et al (eds) Revised selected papers of ICTERI 2013. Communications in computer and information science, vol 412, Springer, Heidelberg, pp 136–162. https://doi.org/10.1007/978-3-319-03998-5_8
6. Kosa V, Chaves-Fraga D, Keberle N, Birukou A (2019) Similar terms grouping yields faster terminological saturation. In: Ermolayev V et al (eds.) Revised selected papers of ICTERI 2018. Communications in computer and information science, vol 1007. Springer, Cham, pp 43–70. https://doi.org/10.1007/978-3-030-13929-2_3
7. Levenshtein VI (1966) Binary codes capable of correcting deletions, insertions, and reversals. Sov Phys Dokl 10(8):707–710
8. Hamming RW (1959) Error detecting and error correcting codes. Bell Syst Tech J 29(2):147–160
9. Jaro MA (1989) Advances in record-linkage methodology as applied to matching the 1985 Census of Tampa, Florida. J Amer Stat Assoc 84(406):414–420. https://doi.org/10.1080/01621459.1989.10478785
10. Winkler WE (1990) String comparator metrics and enhanced decision rules in the Fellegi-Sunter model of record linkage. In: Proceedings of the section on survey research, 1990. ASA, pp 354–359
11. Jaccard P (1912) The distribution of the flora in the alpine zone. New Phytol 11:37–50
12. Yu M, Li G, Deng D, Feng J (2016) String similarity search and join: a survey. Front Comput Sci 10(3):399–417
13. Dice LR (1945) Measures of the amount of ecologic association between species. Ecology 26(3):297–302
14. Sørensen T (1948) A method of establishing groups of equal amplitude in plant sociology based on similarity of species and its application to analyses of the vegetation on Danish commons. Kongelige Danske Videnskabernes Selskab 5(4):1–34

Chapter 5
Experimental Evaluation

This chapter presents the results of the experimental evaluation of the developed method and algorithms, implemented in the software suite. It begins with the outline of our experimental objectives in Sect. 5.1. The explanation of our general experimental set-up is given in Sect. 5.2 and comprises the outline of the experimental workflow and toolset; the description of the document collections used as the input data for the experiments; the presentation of the measurable aspects and measures used in the experiments; and the outline of our experimental environment. The rest of the chapter focuses on the presentation of the specific experiments for evaluating or validating different aspects of the proposed solution for terminological saturation detection and measurement. Section 5.3 presents and discusses our results that prove the correctness of the developed solution in the boundary cases: (i) quick and steady saturation to be observed; and (ii) no saturation to be observed. Section 5.4 elaborates the answer to our research question **RQ3** about the choice of the software tool for ATE. This is done via the cross-evaluation of the two shortlisted tools—NaCTeM TerMine and UPM Term Extractor (c.f. Sect. 2.9). Further, Sect. 5.5 examines the influence of five possible document orderings on the result of terminological saturation measurement. This is done by feeding the datasets, formed using different orderings of documents, into the processing pipeline, and comparing the results using the measurable aspects (Sect. 5.2.3). Section 5.6 focuses on the experimental evaluation of the utility of terms grouping for the developed terminological saturation measurement solution. This evaluation is performed using different string similarity measures and term similarity thresholds. Finally, Sect. 5.7 performs the experimental check of the validity, efficiency, domain neutrality, and scalability of the optimized processing pipeline. Section 5.8 summarizes the findings of our experimental evaluation study.

V. Kosa and V. Ermolayev, *Terminology Saturation*, Cognitive Science and Technology,
https://doi.org/10.1007/978-981-16-8630-6_5

5.1 Experimental Objectives

The objectives of our experiments were framed out by the evaluation and validation facet of the research questions **RQ9**, **3**, **1**, **7**, and **8**, which constituted our research objective **RO3** (c.f. Sect. 2.14, Table 2.5). Specifically, these objectives were as follows:

- **Check if the computation pipeline** for terminological saturation measurement, developed in Chaps. 3 and 4, **produces correct results** both when terminological saturation should be expected very quickly and when it could not be expected. For this check, we prepare two synthetic collections, 1DOC and RAW, in Sect. 5.2.2 and apply the developed processing pipeline to these collections in Sect. 5.3.
- **Recommend the best suitable** external **software implementation** of the ***C*-Value method** to be used for measuring and detecting terminological saturation. For that, we cross-evaluate the two shortlisted software tools, NaCTeM TerMine and UPM Term Extractor (c.f. Sect. 2.9), by the use of these tools in our baseline pipeline applied to three real document collections in Sect. 5.4.
- Find out **which of the possible orderings** of the documents for processing best ensures that terminological saturation is reached with a compact, though representative, retained term-set, quickly, is stable after this point, and is not the saturation of accumulated regular noise. To reach this objective, we cross-evaluate, in Sect. 5.5, five possible orderings and rank the results using the measurable aspects presented in Sect. 5.2.3.
- Evaluate if the use of the **term grouping technique**, based on terms similarity measurements, has a positive impact on the effectiveness and quality of terminological saturation in the terms of faster saturation detection and more significant terms retained. For reaching this objective we cross-evaluate: (i) the baseline computation pipeline using the baseline **`THD`** algorithm; against (ii) the extended pipeline with terms grouping using **`STG`**, **`M-JR`**, **`M-JW`**, **`M-JA`**, **`M-SD`**, and **`R-THD`** algorithms (Sect. 4.6).
- **Validate the optimized method** for terminological saturation measurement (Sect. 3.6). For that, we experimentally prove the $h1$ hypothesis of Theorem 3.5. We also **evaluate the efficiency, domain neutrality, and scalability** of the implemented optimized computation pipeline (Sect. 3.6) by applying its algorithms (Sect. 4.4), implemented in the software suite (Sect. 4.8), to three different real document collections.

5.2 General Experimental Settings

This section explains the general setup of our experimental evaluation. The particularities for individual experimental series, if any, are detailed below in respective Sects. 5.3–5.7. The experimental workflow and instrumental toolset, developed based on the algorithmic suite and its software implementation (Chap. 4), are outlined in

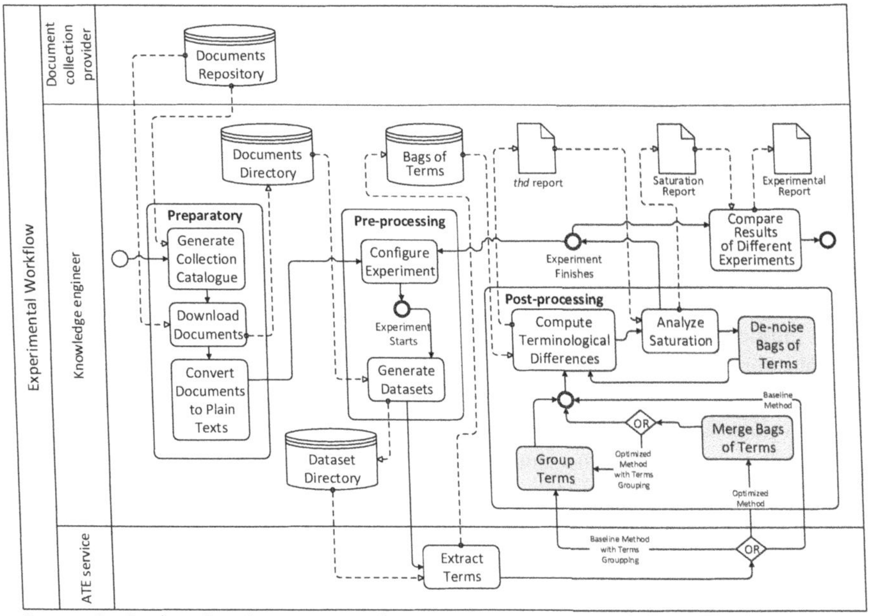

Legend: Optional tasks are greyed

Fig. 5.1 Experimental workflow

Sect. 5.2.1; the document collections and datasets—in Sect. 5.2.2; the measures taken in the experiments—in Sect. 5.2.3; and the experimental environment—in Sect. 5.2.4.

5.2.1 Experimental Workflow and Instrumental Software Toolset

The generic workflow, used in our evaluation experiments, was based on the workflow for terminological saturation detection and measurement (Fig. 4.1). It used the same modules that form the computation flow (Fig. 4.2) and iterated, if necessary, as described in Sect. 4.1. The instrumental software used in the experimental workflow was presented in Sect. 4.8. For the purpose of reference, the experimental workflow is pictured in Fig. 5.1.

5.2.2 Document Collections and Datasets

The data used in our experiments comprised two synthetic and four real document collections.

the proceedings of the Design Automation Conference.[3] Its subject domain was Engineering Design Automation.

For generating the datasets using the TIME, DMKD, and DAC collections, the increment (inc) was chosen to be 20 documents.[4]

Into the DMKD document collection,[5] 300 journal articles, published between 1997 and 2010, were included. The full texts of the papers were automatically processed in the experimental workflow (Fig. 5.1) using our software suite (Chap. 4). 15 incrementally extended datasets, $D_1, D_2, \ldots, D_{15}$, were generated.[6]

The TIME collection has been acquired in our prior research [1]. It was composed of the full-text papers published between 1994 and 2013, which were 437 documents. The documents in TIME were pre-processed manually. Pre-processing included the conversion to and cleaning of the plain texts. Due to manual processing and cleaning, the generated datasets were moderately noisy. 22 incrementally extended datasets $D_1, D_2, \ldots, D_{22}$ were generated.[7]

For the DAC collection, 506 conference papers published between 2004 and 2010 were selected. These documents were automatically converted to plain texts using our instrumental software (Chap. 4). 26 incrementally extended datasets $D_1, D_2, \ldots, D_{26}$[8] were generated. Manual de-noising of the texts was deliberately bypassed to enable the comparison of the results between very noisy bags of terms (without stop terms removal) and not very noisy bags of terms (after stop terms removal) following the approach proposed by Kosa et al. [4].

One may argue that the sizes of the chosen collections and generated datasets were too small to provide sufficient and valid experimental evidence. This was not the case, however. All the three document collections were sampled from the bigger collections in a way to have them as small as possible to observe terminological saturation. In the presented research, the smaller was indeed the better as the differences in the measurable aspects (Sects. 5.4–5.6) were better visible with smaller datasets and their increments. Furthermore, the scalability and computational efficiency of the instrumental pipeline for terminological saturation measurement has

---

[3] http://dac.com/.

[4] 20 documents were considered as an appropriately granular increment for the diagrams and tables representing experimental results. According to the Corollary 3.6 (c.f. [3]), the size of the increment did not influence the result.

[5] This collection was provided by Springer-Nature based on their policy on full text provision for data mining purposes: https://www.springer.com/gp/rights-permissions/springer-s-text-and-data-mining-policy/29056.

[6] **DMKD** collection in plain texts: http://dx.doi.org/10.17632/knb8fgyr8n.1#folder-637dc34c-fa29-4587-9f63-df0e602d6e86; incrementally extended datasets generated of these texts: http://dx.doi.org/10.17632/knb8fgyr8n.1#folder-9043e879-5f77-486f-bc56-cb6af3cdd306.

[7] **TIME** collection in plain texts: http://dx.doi.org/10.17632/knb8fgyr8n.1#folder-d1e5f2b6-c51e-4572-b10d-0e2ebccead02; incrementally extended datasets generated of these texts: http://dx.doi.org/10.17632/knb8fgyr8n.1#folder-06cc93de-de58-43b9-a378-743b61ef9093.

[8] **DAC** collection in plain texts: http://dx.doi.org/10.17632/knb8fgyr8n.1#folder-010b1add-cd5c-4b33-b6ce-8c93301d880b; incrementally extended datasets generated of these texts: http://dx.doi.org/10.17632/knb8fgyr8n.1#folder-e07d55cc-6830-4d9e-9b90-d69010050508.

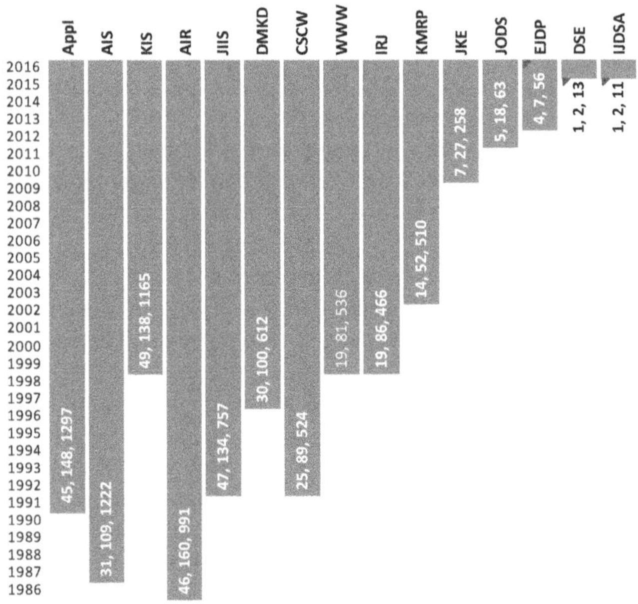

Legend: Y - axis represents the years of publication and X-axis corresponds to the journals (acronyms are given). The numbers in the bars are the number of volumes, of issues, and the total number of papers in the journal.

Fig. 5.2 The distribution of articles in the journals used as the sources for the KM collection adapted from [2]

been tested, using the **dcf** ordering [3], on a complete KM document collection [2] in our optimization and scalability experiments (Sect. 5.7).

The KM collection contained over 9000 articles, belonging to 15 Springer-Nature journals, focusing on different sub-fields in Knowledge Management. The constitution of the collection is pictured in Fig. 5.2.

Some of the articles in KM were considered insufficiently informative for extracting terms. These articles were editorials and the like, and the articles without full texts. After filtering out these documents, the collection contained 7445 full-text journal articles.[9] These documents were automatically converted to plain texts [4] and were not cleaned. Therefore, the resulting datasets were moderately noisy, as in the DMKD collection. For generating the KM datasets, the increment of inc =

[9] The full texts were provided by Springer-Nature based on their policy on full text provision for data mining purposes: https://www.springer.com/gp/rights-permissions/springer-s-text-and-data-mining-policy/29056. The volume of the KM collection after conversion to plain texts is 413.66 Mb.

Table 5.1 The summary of the characteristics of document collections and datasets

Collection	Type	Document type and layout	No. of documents	Noise	Increment (no. of documents)	No. of datasets
1DOC	Synthetic	Journal, ACM 1-column	1	Manually cleaned	1 paper	20
RAW	Synthetic	Wikipedia 1-column	80	Not cleaned, moderately noisy	4 papers	20
DMKD	Real	Journal, Springer 1-column	300	Not cleaned, moderately noisy, regular noise not accumulated	20 papers	15
TIME	Real	Conference, IEEE 2-column	437	Manually processed, moderately noisy, regular noise accumulated	20 papers	22
DAC	Real	Conference, IEEE 2-column	506	Not cleaned, substantially noisy, regular noise accumulated	20 papers	26
KM	Real	Journal, Springer 1-column	7445	Not cleaned, moderately noisy, regular noise not accumulated	100 papers	75

100 documents[10] were chosen. As a result, 75 datasets of inc size, forming the partition $\left\{D_i = \{d_j\}_{j=1}^{100}\right\}_{i=1}^{75}$ of the KM collection,[11] were generated for the optimized processing pipeline. Incrementally extended datasets were not generated.

The characteristics of all the document collections and datasets used in our experiments are summarized in Table 5.1.

[10] 100 documents were considered as an appropriately granular increment for the diagrams and tables representing experimental results.

[11] KM partition datasets have not been made publicly available, as it requires additional permissions by the owner of the data.

5.2.3 Measurable Aspects and Measures

Our experiments studied the influences of different aspects of the used tools and data on: (i) terminological saturation; (ii) accumulated regular noise detection; (iii) method performance, including its scalability. Table 5.2 presents the measurable aspects used in the analysis of the experimental results, with regard to our objectives (Sect. 5.1).

To evaluate these aspects, the following measures were taken in different experiments:

- NET—the number of extracted tern candidates (in B_i),
- *eps*—individual term significance threshold,
- NRT—the number of retained significant terms (in T_i),
- PRT—the proportion of the numbers of retained significant (NRT) to all extracted (NET) terms in percent,
- *thd*—the value of terminological difference between T_i and T_{i-1},
- *thdr*—the value of normalized terminological difference between T_i and T_{i-1},
- ThdV—a discrete analogue of the 1st derivative to thd values computed as the difference between the current $thd(T_i, T_{i-1})$ and previous $thd(T_{i-1}, T_{i-2})$ values.

5.2.4 Experimental Environment

All the computations, except terms extraction, were executed using a Windows 7 64-bit PC with Intel® Core™ i5 CPU, M520 @ 2.40 GHz; 8.0 Gb onboard memory; NVIDIA Geforce GT330M GPU. For terms extraction, an Ubuntu 16.04LTS PC with Intel® Xeon® CPU E5-2603 v3 @ 1.60 GHz 12 cores, 128G on-board memory, and 3.5 T hard disk was used.

5.3 Correctness Check Using Synthetic Collections

The objective of this experimental series was to validate the correctness of the instrumental software pipeline, including both shortlisted tools for ATE (NaCTeM TerMine and UPM Term Extractor, Sect. 2.9) with regard to the boundary cases: (i) 1DOC collection; and (ii) RAW collection. Due to the design of these synthetic collections (Sect. 5.2.2), it was expected that the results on 1DOC demonstrated immediate and steady terminological saturation, while the results on RAW highlighted the absence of terminological saturation.

Table 5.2 Measurable aspects used in the experiments

Aspect	Label	What is measured	Rationale	Ranking rule
The study of the influence on terminological saturation (DMKD, TIME, DAC-cleaned datasets)				
(i) Compactness	*Seps*	Individual term significance threshold (*eps*) value at the saturation point	It implies from the definition of *eps* in Sect. 3.1 (Definition 3.4) that the higher values result in more compact sets of retained significant terms (simple majority principle)	The higher–the better
(ii) Performance	*SthdSP*	Point of entry into the saturation zone (*thd* lower than *eps*)—saturation point	The earlier terminological saturation is detected the less measurements need to be performed on the smaller dimension datasets	The earlier–the better
(iii) Representativeness	*SPrtSP*	The average proportion of the numbers of retained significant terms to all extracted terms throughout the entire range of measurements	A higher share of retained significant terms straightforwardly indicates a higher recall, hence results in a better representative term-set. An average based on the entire measurement interval is used to bypass fluctuations at individual points	The higher–the better
(iv) Stability	*SthdV*	The integral volatility of thd values in the saturation zone (from the point of entry till the last measurement not influenced by regular noise accumulation)	The lower the observed volatility of thd values is integral, the less probable is that terminological saturation is broken in the subsequent measurement points—hence, it is more stable	The lower–the better

(continued)

Table 5.2 (continued)

Aspect	Label	What is measured	Rationale	Ranking rule
The study of the sensitivity to excessive regular noise (DAC Naturelle collection)				
(v) Sensitivity to accumulated regular noise	*NepsPH*	The height of the peak of *eps*	If a sharp and high peak of *eps* is observed, then it might be supposed that a regular noise (with high *C*-Values) has been accumulated. As these regular noise items are accumulated in the upper part of a bag of terms, there is no more room for real term candidates. Hence, any further extractions are useless as those are the extractions of regular noise	The higher–the better
	NepsPP	The measurement point of the peak of *eps*		The earlier–the better
	NthdPP	The measurement point of the peak of *thd*	Based on the definition of *thd* as a terminological difference measure, it might be expected that, in the presence of accumulated regular noise, it would: (a) oscillate sharply with high amplitude; and (b) drop down to near to zero values shortly after	The earlier—the better

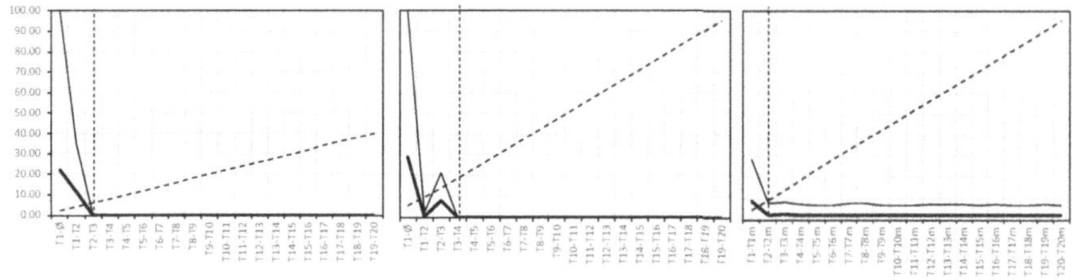

(a) Result for NaCTeM TerMine (b) Result for UPM Term Extractor (c) Difference between (a) and (b)

Legend: ----- *eps*, — *thd*, — *thdr*

Fig. 5.3 Comparison of terminological saturation measurements for the 1DOC datasets

5.3.1 *Results and Discussion*

With respect to the 1DOC datasets and the bags of terms extracted from these datasets, the results were as follows.

The bags of terms extracted by NaCTeM TerMine were processed first. The results of measuring individual term significance thresholds (*eps*) and terminological differences (*thd*, *thdr*) are pictured in Fig. 5.3a.[12] Further, the terminological differences between the bags of terms extracted by UPM Term Extractor were measured. The results of measuring individual term significance thresholds (*eps*) and terminological differences (*thd*, *thdr*) are shown in Fig. 5.3b.

In the case of using NaCTeM TerMine for terms extraction, the dashed vertical line in Fig. 5.3a points to the iteration ($T_3 - T_2$) in which terminological saturation was detected for the first time as *thd* went below *eps*. As was expected, we observed steady saturation in all subsequent iterations. Saturation measurements showed the same number of significant retained terms and increase in the values of individual term significance threshold (*eps*). The values of *thd* and *thdr* went down to the proximity of zero starting from ($T_3 - T_2$).

In the case of using UPM Term Extractor for terms extraction, the dashed vertical line in Fig. 5.3b points to the iteration ($T_2 - T_1$) in which terminological saturation was for the first time as *thd* went below *eps*. Very similar to the case of NaCTeM TerMine, we observed stable terminological saturation in all subsequent iterations. The same numbers of extracted terms and increase in individual term significance threshold values (*eps*) were observed. The values of *thd* and *thdr* went down to almost zero starting from ($T_4 - T_3$).

Some differences in terminological saturation measurements were noticed however for the bags of retained significant terms extracted using NaCTeM TerMine and UPM Term Extractor:

[12] For the sake of making the discussion more concise, the full details of measurements are omitted. Some of the values are mentioned for illustrative purposes. The full details on these measurements are presented in the technical report [5], which is publicly available online.

(i) UPM Term Extractor extracted bigger bags of term candidate strings with C-Value > 1: 3019 term candidates versus 1208 extracted using NaCTeM TerMine.
(ii) Individual term significance threshold values (*eps*) were about 2.5 times higher for the bags of terms-candidates extracted by UPM Term Extractor.
(iii) The number of retained significant terms, with C-Value > *eps*, was circa two times bigger in the UPM Term Extractor case.

Overall, compared to UPM Term Extractor results, NaCTeM TerMine results showed a slightly quicker convergence to steady terminological saturation. On the other hand:

(i) The number of retained significant terms at the saturation point; and
(ii) The individual term significance threshold at the saturation point

were higher in the UPM Term Extractor results. These differences, allowed us to conclude that, in the linguistic pipeline, NaCTeM TerMine was approximately three times more selective with regard to the extraction of candidate terms. This implied that pre-processing in NaCTeM TerMine was more sophisticated. On the other hand, the terms, retained from those extracted by UPM Term Extractor, appeared to be approximately two times more significant at the cut off. Therefore, it could have been noted that the statistical part in UPM Term Extractor processing pipeline extracted more compact, yet significant sets of terms. These observations allowed concluding that, due to the statistical processing phase, UPM Term Extractor was a more selective tool.

It was further checked if both evaluated tools extracted similar bags of terms from 1DOC. For checking, terminological differences between the bags of retained significant terms, coming from NaCTeM TerMine (T_i) and UPM Term Extractor (T_i^m), were measured for every iteration i, as pictured in Fig. 5.3c. The figure shows that both tools extracted statistically very similar bags of retained significant terms despite the differences in the numbers of retained terms and *eps* values. In these comparisons, the values of $thd(T_i, T_i^m)$ became negligible at the second iteration ($T_2 - T_2^m$) as the *thd* value (2.291409) went times below the *eps* value (9.509775). In the subsequent iterations, the *thd* values oscillated around 2.1 and the *eps* values steadily increased up to 95.

With respect to the RAW datasets and the bags of terms extracted from these datasets, the results were as follows.

The bags of terms extracted by NaCTeM TerMine were processed first. The results of measuring individual term significance thresholds (*eps*) and terminological differences (*thd*, *thdr*) are shown in Fig. 5.4a.

The biggest bag of candidate terms (B_{20}) extracted by NaCTeM TerMine was further manually checked from the top down to the terms having C-Values higher than 40. This scan yielded a list of approximately two hundred stop terms. The stop terms were removed from $B_1, \ldots, B_{20}$ and saturation measurements were repeated. The results for the de-noised bags of term candidates are presented in Fig. 5.4b.

When comparing Fig. 5.4a and b, one could observe that, as expected, RAW was not terminologically saturated. A closer check of the differences between Fig. 5.4a

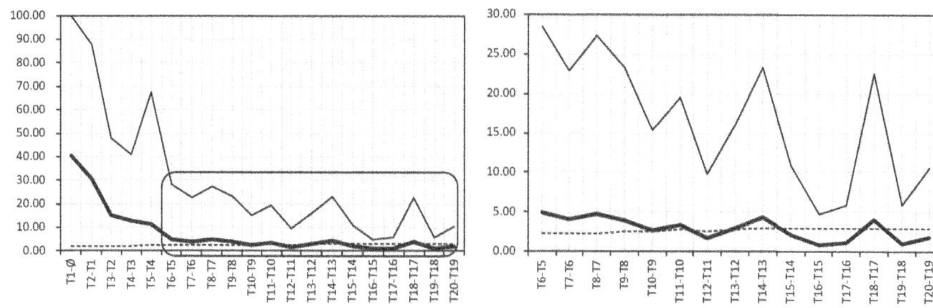

(a) **Before** removing regular accumulated noise

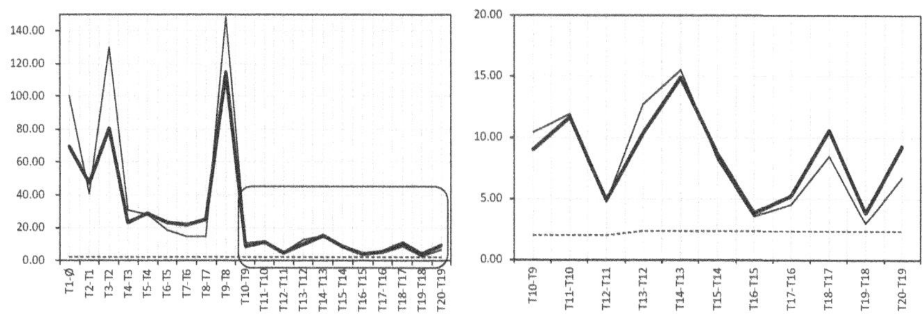

(b) **After** removing regular accumulated noise

Legend: ------ *eps*, ▬▬ *thd*, —— *thdr*; the diagram at the right represents a more granular look into the rounded rectangle in the diagram at the left

Fig. 5.4 Terminological saturation measurements for the RAW bags of terms extracted by NaCTeM TerMine

and b reveal some indicators of the presence of accumulated regular noise in the datasets. The values of *thdr* in Fig. 5.4a were much bigger than the *thd* values. Though the values of *thd* indicated that the bags of retained significant terms were close to saturation, the values of *thdr* were significantly higher than the values of *eps*. It is worth further notice, that the *thd* and *thdr* values, measured after stop-term removals, became very similar. Furthermore, the *thd* and *thdr* curves in Fig. 5.4b became very similar to the *thdr* curve in Fig. 5.4a. Hence, substantial differences between *thd* and *thdr* values might be considered an indicator of excessive regular accumulated noise that needs to be removed from the bags of term candidates.

The same experiment has been then executed for the bags of terms extracted by the UPM Term Extractor. The results were as follows.

The results of measuring individual term significance thresholds (*eps*) and terminological differences (*thd*, *thdr*) are shown in Fig. 5.5a. Similar to the previous experiment that used NaCTeM TerMine, B_{20} extracted by UPM Term Extractor was manually checked for accumulated regular noise from the top down to the term candidates with *C*-Values bigger than 40. This scan yielded two hundred and 20 stop terms. The stop terms have been removed from $B_1, \ldots, B_{20}$ and saturation measurements

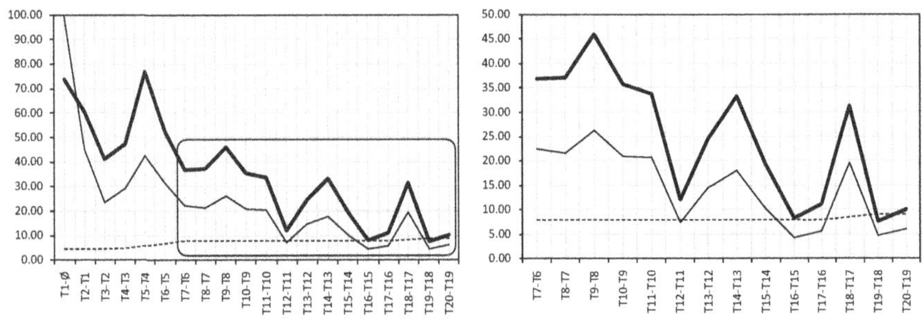

(a) **Before** removing regular accumulated noise

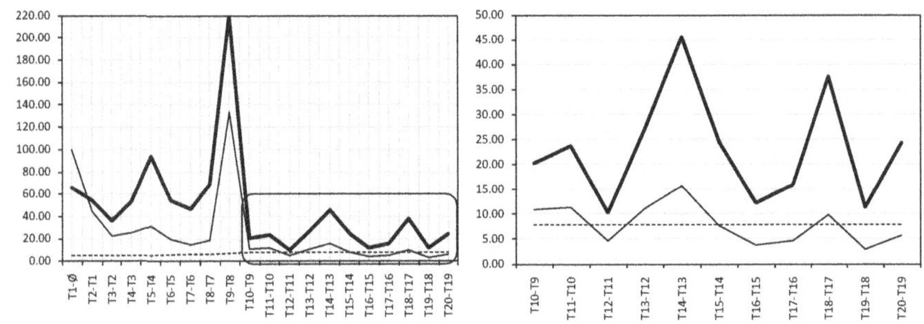

(b) **After** removing regular accumulated noise

Legend: ----- *eps*, ━━ *thd*, —— *thdr*; the diagram at the right represents a more granular look into the rounded rectangle in the diagram at the left

Fig. 5.5 Terminological saturation measurements for the RAW bags of terms extracted by UPM Term Extractor

were repeated. The results for the de-noised bags of term candidates are pictured in Fig. 5.5b.

Compared to the terminological saturation measurements for the bags of terms extracted by NaCTeM TerMine, the values of *thd* for the bags of terms extracted by UPM Term Extractor present a clearer picture of the absence of terminological saturation. The *thd* values, measured before removing the stop terms from the UPM Term Extractor results, were 2.5–3 times higher than the measures on NaCTeM TerMine results after removing the stop terms. Hence, the results by UPM Term Extractor were more contrasting, compared to those of NaCTeM TerMine, for detecting the absence of terminological saturation. On the other hand, the values of *thdr* measured for NaCTeM TerMine outputs provided a sharper indicator of the need to de-noise the bags of term candidates. The *thdr* values measured for the UPM Term Extractor outputs did not substantially differ from the corresponding *thd* values. However, if UPM Term Extractor was used to detect the absence of terminological saturation, there was no need to analyse if *thdr* values indicated the presence of noise, as it was irrelevant. Hence, the use of UPM Term Extractor is preferred in this case as it is a more sensitive tool.

In the case of RAW, it was not checked if both tools extracted similar bags of term candidates. This check would have had no value in the absence of terminological saturation.

5.3.2 *Recommendation of the ATE Tool*

Based on the results of our correctness check experiments (Sect. 5.3.1), it may be concluded that:

(i) Both evaluated tools performed correctly and delivered the results that have been expected; and
(ii) The use of UPM Term Extractor is preferred, over NaCTeM TerMine, both for terminologically saturated document collections and detecting that saturation could hardly be expected.

5.4 Choice of Software for ATE

This Section reports our results[13] of the further cross-evaluation of the UPM Term Extractor and NaCTeM TerMine using the real document collections, DMKD, TIME, and DAC (Sect. 5.2.2).

5.4.1 *Results of Experiments*

For the DMKD collection, the results were as follows. The bags of terms extracted by NaCTeM TerMine were processed first. After that, the experiment was repeated using UPM Term Extractor.

Regarding the NaCTeM TerMine case, the results of measuring individual term significance thresholds (*eps*) and terminological differences (*thd*, *thdr*) are pictured in Fig. 5.6. The diagram at the left reports the measures for all the iterations that were performed. The rounded rectangular in this diagram shows the subset of measurements, in the saturation proximity zone, that is pictured in finer detail in the diagram at the right. The dashed vertical line marks the saturation point at the iteration ($T_{14}-T_{13}$), when *thd* went below *eps*.

These results indicate the trend of reaching stable terminological saturation if more documents were available in the collection. The values of *eps* were increasing and the values of *thd* and *thdr* demonstrated the tendency to go down, with some volatility that had decreasing amplitudes with the growth of iteration numbers. The increments in the numbers of retained significant terms were also decreasing. There were

[13] The material of this section is based on our publication [4].

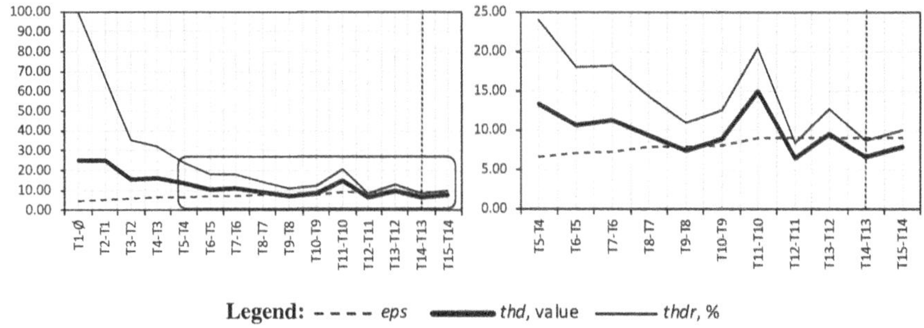

Fig. 5.6 Terminological saturation measurements for the bags of term candidates extracted by NaCTeM TerMine from DMKD datasets

three terminological peaks observed in the saturation area at $(T_{11} - T_{10})$, $(T_{13} - T_{12})$, and $(T_{15} - T_{14})$. The peaks indicated the inflows of more new significant terms at these iterations. These inflows were not substantial as the values of *thd* did not increase very much compared to the vicinity (c.f. the contrasting results for DAC in Fig. 5.10). Based on these observations, it was premature to consider DMKD a saturated document collection if term candidates were extracted by NaCTeM TerMine.

The results of measuring individual term significance thresholds (*eps*) and terminological differences (*thd*, *thdr*) for the bags of terms extracted by UPM Term Extractor are pictured in Fig. 5.7. For this case, it was observed that steady saturation was reached at $(T_6 - T_5)$. The number of significant terms retained from B_6 was 4,113, which is noticeably less (about 18%) than the number of retained significant terms (5,009) at the first potential saturation point $(T_9 - T_8)$ in the NaCTeM TerMine case—c.f. Fig. 5.6.

It is worth noting that *thd* and *thdr* values in the UPM Term Extractor case demonstrated the fluctuations that resembled the results in the NaCTeM TerMine case. These values hinted at terminological peaks at the same iterations. The numbers of retained significant terms in the UPM Term Extractor case were lower, though not

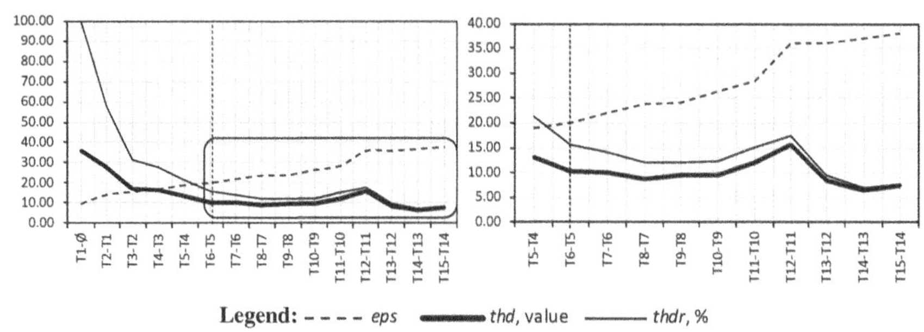

Fig. 5.7 Terminological saturation measurements for the bags of term candidates extracted by UPM Term Extractor from DMKD datasets

substantially. Terminological saturation was reached due to much higher values of *eps*.

In a summary for DMKD, it could be stated that the use of UPM Term Extractor resulted in extracting compacter bags of retained significant terms. Furthermore, the insignificant terms were cut off at the higher values of individual term significance threshold.

Our hypothesis about the reason why UPM Term Extractor performed better was that it extracted not all the term candidates from the input datasets, in contrast to the use of NaCTeM TerMine, which yielded a higher recall. To check that, terminological differences between the bags of terms extracted, from the same datasets by UPM Term Extractor and NaCTeM TerMine were measured. The result is visualized in Fig. 5.8.

Figure 5.8 shows that both tools extracted similar bags of term candidates. Terminological difference measures decreased with the growth of compared datasets in iterations. These indicated that the bags of retained significant terms, extracted using the compared tools, were similar, hence the recalls achieved by the tools differed not too much.

Interestingly, *thd* in Fig. 5.8 went below *eps* exactly at the point when the *thd* measurements in the NaCTeM TerMine case (c.f. Fig. 5.6) reveal the highest terminological peak. It looked like both tools allowed extracting similar bags of retained significant terms but NaCTeM TerMine yielded saturated term-sets later than UPM Term Extractor when it collected the inputs from the documents contributing to the highest terminological peak.

Yet interestingly, thd values went above the values of *eps* after $(T_{11} - T_{11}^{m})$. A possible reason for that might be the accumulation of regular noise in the datasets, which is processed differently by the compared tools.

The results of the cross-evaluation of NaCTeM TerMine and UPM Term Extractor using the TIME datasets are visualized in Fig. 5.9.

As pictured in Fig. 5.9a, the measurements for the bags of term candidate strings extracted by NaCTeM TerMine revealed the absence of terminological saturation. The values of *thd* and *thdr* did not go below the *eps* curve. The shapes of *thd* and

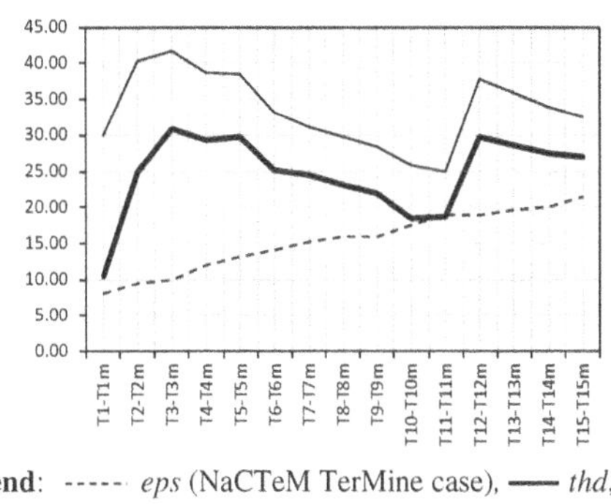

Fig. 5.8 Terminological differences between the bags of terms extracted from DMKD by UPM Term Extractor and NaCTeM TerMine

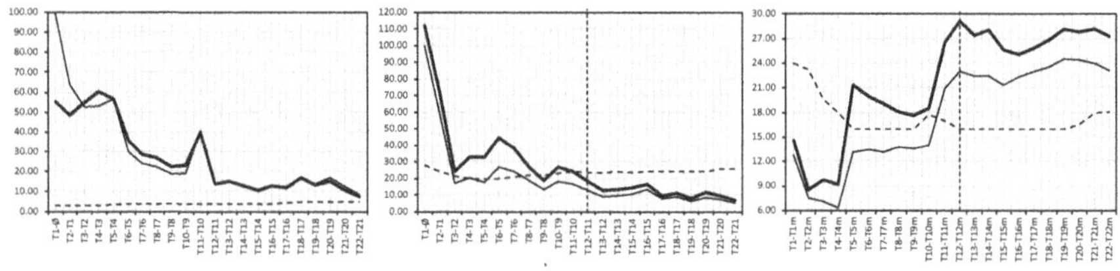

(a) NaCTeM TerMine (b) UPM Term Extractor (c) Difference between (a) and (b)

Legend: - - - *eps* ▬▬ *thd*, value —— *thdr*, %

Fig. 5.9 Terminological saturation measurements for the TIME collection

thdr curves were quite similar to those observed for DMKD indicating the same trend—potential terminological saturation if more documents were available in the collection. The values of *eps* also went up in iterations, though noticeably slower than in the DMKD case. The maximal *eps* value was 5 versus 9 for DMKD. Though the values of *thd* and *thdr* went down in iterations, but not sufficiently quickly to reach saturation. The maximal number of retained significant terms for TIME was significantly higher than in the DMKD case: 8,343 against 5,438. However, the difference in the numbers of extracted term candidate strings was not that high: ~ 287 K for TIME against ~ 253 K for DMKD. Terminological peaks were observed at $(T_4 - T_3)$, $(T_{11} - T_{10})$, $(T_{18} - T_{17})$, and $(T_{20} - T_{19})$ with the highest at $(T_{11} - T_{10})$. As in the case of DMKD, terminological contributions of these peaks were not a game changer because the increases in *thd* values did not change the general trend of decrease with iterations.

In difference to the NaCTeM TerMine case, the measurements for the bags of term candidates extracted by UPM Term Extractor revealed terminological saturation starting from $(T_{12} - T_{11})$, as demonstrated in Fig. 5.9b—the saturation point is marked by the vertical dashed line. The values of *thd* and *thdr* in these measurement series resembled the NaCTeM TerMine case quite closely. In particular, *thd* and *thdr* curves peaked nearly at the same iterations, though with lower amplitudes. The *eps* values were, however, much higher, as in DMKD experiment. Terminological saturation was detected at $eps = 23.774$, whereas in the NaCTeM TerMine case *eps* values were not higher than 5.0. The number of significant terms retained in T_{12} was 7,110. This number was only 2.47% of the total number of the term candidate strings in B_{12}.

Based on the analysis of the findings in the cross-evaluation of the tools using TIME, a conclusion similar to that of the DMKD case could be drawn. Terminological saturation in TIME was reached using the bags of term candidates extracted by UPM Term Extractor because the values of the individual term significance threshold (*eps*) appeared to be much higher than in the NaCTeM TerMine case. Similar to DMKD, the use of UPM Term Extractor allowed retaining compacter bags of significant terms.

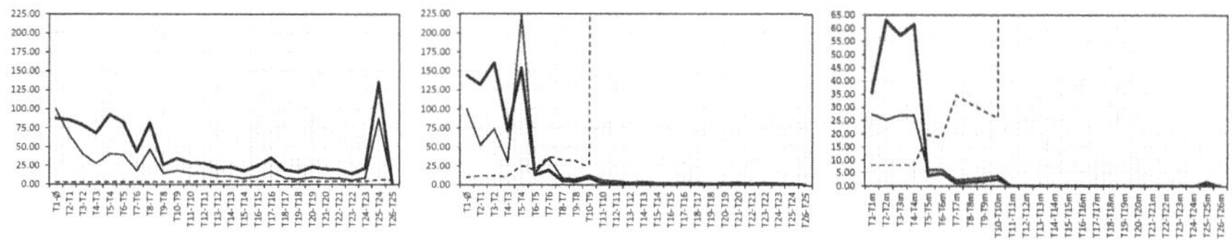

(a) NaCTeM TerMine (b) UPM Term Extractor (c) Difference between (a) and (b)

Legend: ---- *eps* —— *thd*, value —— *thdr*, %

Fig. 5.10 Terminological saturation measurements for the DAC collection

It was also checked if the cross-evaluated tools extracted similar bags of terms from the TIME collection. The measurements (Fig. 5.9c) were done following the same approach as in the DMKD case. The diagram shows that the terminological difference (*thd*) between the bags of retained significant terms at the saturation point $\left(T_{12} - T_{12}^{m}\right)$[14] was circa 29, which was 1.81 times higher than $eps = 16$. In the DMKD case, the proportion of *thd* to *eps* at the saturation point was almost the same—1.80. As in the case of DMKD, $thd\left(T_i, T_i^{m}\right)$ for TIME grew in the iterations subsequent to the saturation point, probably due to the same reason—the accumulation of regular noise.

The results of the cross-evaluation of NaCTeM TerMine and UPM Term Extractor using DAC datasets are shown in Fig. 5.10. DAC documents contained much more regular noise than DMKD or TIME as the text files were deliberately left uncleaned to investigate the impact of the accumulated regular noise on terminological saturation. Therefore, the results of saturation measurements were different both in values and in observed trends.

The measurements for the bags of term candidates extracted by NaCTeM TerMine indicated potential saturation only in the last iteration $(T_{26} - T_{25})$, as pictured in Fig. 5.10a. However, the terminological peak in the previous iteration $(T_{25} - T_{24})$, with $thd = 135.49$, hinted that, in the subsequent iterations, the convergence to saturation would have not been observed with sufficient stability. Hence, more measurements, using additional collection increments, were required for a justified conclusion.

It was also interesting to compare saturation trend for DAC to that for TIME, due to the similarities in document layouts and sizes. The datasets might have been different in the levels of accumulated regular noise only because TIME documents were manually cleaned after the conversion to plain texts, while DAC documents were deliberately not cleaned. The differences in *thd*, *thdr*, and *eps* values for the datasets of similar sizes could be observed via the comparison of Fig. 5.9a for TIME and Fig. 5.10a for DAC. This comparison reveals that:

---

[14] T_{12} is the bag of significant terms retained from B_{12}. B_{12} is the bag of term candidate strings extracted from the dataset D_{12} using UPM Term Extractor. T_{12}^{m} is the bag of significant terms computed using the similar pipeline using NaCTeM TerMine as term candidates extractor.

(i) The values of individual term significance threshold (*eps*) grew faster in iterations for the TIME datasets than for the DAC datasets.
(ii) Much more term candidates were extracted by NaCTeM TerMine and significant terms were retained from the DAC datasets than from the TIME datasets.
(iii) In the case of TIME, the number of retained significant terms grew monotonically. This growth slowed down in iterations, which indicated an approach to possible saturation in subsequent iterations.
(iv) In difference to the case of TIME, the number of significant terms retained from DAC datasets dropped significantly below the value of the previous iteration at $(T_{24} - T_{23})$. In addition to that, a sharp increase of *thd* was observed: from $thd(T_{24} - T_{23}) = 21.51$ to $thd(T_{25} - T_{24}) = 135.49$. These changes in *eps* and *thd* were caused by the accumulation of regular noise. This was verified by the manual inspection of the corresponding bags of term candidate strings and confirmed by a substantial decrease in the number of retained significant terms.[15] NaCTeM TerMine signalled about the presence of noise, but it happened quite lately.

The measurements for the bags of term candidates extracted by UPM Term Extractor revealed steady terminological saturation starting at $(T_6 - T_5)$. At this point, *eps* value was about 20, as visualized in Fig. 5.10b; *thd* and *thdr* curves went below the curve of *eps*. In the subsequent iterations, however, *eps* values grew up to the peak of 18,294 at $(T_{11} - T_{10})$. Simultaneously, the numbers of retained significant terms went down to 34, which is more than 100 times lower than at the previous iteration $(T_{10} - T_9)$. The manual inspection of the bag of retained significant terms T_{10} revealed that these 34 retained terms were all accumulated regular noise strings. Hence, UPM Term Extractor alerted about the accumulation of regular noise significantly earlier than NaCTeM TerMine. Therefore, the observed sensitivity of UPM Term Extractor to the accumulation of regular noise was noticeably higher than the sensitivity of NaCTeM TerMine.

Furthermore, the behaviour of *eps* and *thd* values, observed in this experiment (Fig. 5.10b), could have been regarded as a signal of the need to remove the accumulated regular noise from the extracted bags of term candidate strings.

The similarity of the bags of term candidates extracted by both cross-evaluated tools was also checked in the DAC case. The result of this check is visualized in Fig. 5.10c. Manual inspection revealed that, starting from $(T_5 - T_5^m)$, with $thd = 3.97$ and $eps = 19.65$, both tools extracted very similar bags of accumulated noise strings.

[15] The inspection of the bags of term candidate strings (B_{25}) and retained significant terms (T_{25}) revealed that the majority of the retained terms were, in fact, regular noise strings.

5.4.2 *Discussion and Recommendation*

For DMKD, which has been automatically pre-processed (Sect. 5.2.2), it could not have been reliably judged that the DMKD collection was terminologically saturated based on the terms extraction using NaCTeM TerMine. In difference to that, terminological saturation measurements using the bags of terms extracted by UPM Term Extractor revealed steady terminological saturation quite quickly. Both tools extracted similar bags of terms.

In the case of TIME, which has been pre-processed and de-noised manually, terminological saturation measurements, using the bags of terms extracted by NaCTeM TerMine, failed to detect saturation. Very similarly to the DMKD case, terminological saturation measurements using the bags of terms extracted by UPM Term Extractor detected steady terminological saturation quite quickly, also with much higher individual term significance thresholds (*eps*). The latter also contributed to extracting the compacter sets of retained significant terms.

The cases of DMKD and TIME demonstrated similar advantages of UPM Term Extractor over NaCTeM TerMine in detecting terminological saturation and retaining significant terms. UPM Term Extractor yielded the compacter sets of retained significant terms. Manual de-noising of the TIME collection did not help improve the results of terminological saturation measurements.

For DAC, which has been deliberately left very noisy, UPM Term Extractor demonstrated the sensitivity sufficient to detect the accumulation of excessive regular noise in the bags of terms much earlier than NaCTeM TerMine. Terminological saturation curve, built for the measurements using UPM Term Extractor results, signalled the presence of excessive regular noise with much higher amplitude. At the same time, the numbers of retained significant terms dropped down by two orders of magnitude, and individual term significance thresholds peaked up by three orders of magnitude. Hence, in the case of noisy datasets, UPM Term Extractor is much more sensitive in detecting accumulated regular noise than NaCTeM TerMine.

The summary of the observations in this series of experiments allows concluding that the use of UPM Term Extractor is preferred over NaCTeM TerMine to detect terminological saturation or excessive regular noise. This recommendation is domain neutral.

5.5 Influence of Document Ordering

This section presents and discusses the results of our experiments aimed at examining what ordering of documents for the inclusion in the datasets is the most preferable. This presentation commences with the outline of the additional experimental settings in the context of this objective (Sect. 5.5.1). The discussion continues with the results of the comparative study of the influence of different possible orderings on terminological saturation in Sect. 5.5.2. Further, the results of the study of the sensitivity to regular noise accumulation are discussed in Sect. 5.5.3. In Sect. 5.5.4, order rankings based on the two studies are given and our recommendation of the ordering is drawn.

All the experimental results are presented in full detail, as tables and charts, in the supporting technical report [6][16] and are available as a part of our public OntoElect Dataset.[17]

5.5.1 Particularities in Experimental Settings

It has been outlined in Sect. 3.1 that a successive approximation approach for building the sequence of document sub-collections $DSC_1, \ldots, DSC_i, \ldots$, was incremental. Therefore, $DSC_1 = INC_1$ and $\forall i \geq 2, DSC_i = DSC_{i-1} \cup INC_i$, where INC_i was a subset of a certain number (*inc*) of documents from a *CDC*. Several reservations need to be made with regard to this iterative process.

(i) **Compactness**. Adding new documents to the processed sub-collection implies that the number of extracted terms grows with the number of iterations. Similarly, the number of retained significant terms grows noticeably if the individual term significance threshold eps_i does not grow sufficiently quickly. The higher the number of retained terms, the more expensive is the subsequent part of an ontology learning pipeline. Hence, our objective is to find a way to maximize the growth of *eps* to have more compact yet representative sets of retained significant terms.

(ii) **Performance**. The incremental growth of sub-collections implies that the more iterations are undertaken, the more computationally expensive such a pipeline is. Hence, one of the objectives of our research is to find a way to reach terminological saturation as quickly as possible to lower the incurred computational overhead.[18]

(iii) **Representativeness**. In its essence, our approach (Sect. 3.1) is based on balancing at the edge of a simple majority vote regarding the computation of *eps* at every iteration. Hence, the sets of the retained significant terms cannot possess high terminological recall. Finding a way to increase the proportions of retained significance to all extracted terms, while maintaining the compactness of the retained term-sets, may therefore increase their representativeness and the quality of term extraction. Improving representativeness is one more objective in our research.

(iv) **Stability**. As it has been demonstrated in our prior research, e.g. [1], terminological saturation condition (3.4) may be violated after it has been reached

[16] Technical report [6] is publicly available at https://github.com/OntoElect/Docs/blob/master/Reports/TS-RTDC-TR-2018-2-v2.pdf.

[17] http://dx.doi.org/10.17632/knb8fgyr8n.1#folder-348f6201-7a55-4c96-a67a-47ec54c9d558.

[18] An alternative way to increase performance is to use document collection partitioning and compute Merged Partial *C*-Values [3] instead of *C*-Values [7]. Anyhow, the approach of [3] is also iterative. Hence, the earlier terminological saturation is detected the better is the performance.

in previous iterations. A reason for that might be the income of a substantial quantity of new significant terms in subsequent iterations. In such cases, terminological saturation is qualified as unstable. Consequently, one more of our objectives is to find a way to minimize the oddness in retained significant terms distributions in the sequence of incrementally extended sub-collections to improve the stability of terminological saturation.

(v) **Sensitivity to accumulated regular noise**. Due to the incremental and iterative nature of our pipeline, the fragments of the texts, which are regular noise items (e.g. a frequently mentioned publication venue or a string of dashes representing a line), might be accumulated with high *score*s. These stop terms may occupy the upper parts of the bags of retained significant terms. Hence, terminological saturation that is detected is in fact the saturation of the stop terms. To avoid these false positives, a way is sought to increase the sensitivity of the pipeline to accumulated regular noise.

In a summarizing manner, these aspects (i–v) were, in fact, put together in the research question (**RQ1**—document ordering, Sect. 2.13.2). Here, the formulation of this research question has to be extended to expose better how the aspects (i–v) are taken into a balanced account:

RQ1 (extended): *How to order the documents from a CDC for forming INC_i to ensure that terminological saturation is reached with a compact though representative retained term-set, quickly, is stable after this point, and is not the saturation of accumulated regular noise?*

To answer this question, we presume that the collections are composed of relevant documents published at different times. Therefore, it is possible, for the domains evolving in time, that the constituent documents contain different terms and the distribution of these terms is not even throughout the collection. This phenomenon is termed **temporal drift** in terminology. Some of the terms may pass the test of time and remain significant for describing the domain for long. Hence, these terms are spread quite evenly throughout the entire collection. Other terms may become marginally significant quite quickly after their introduction. Hence, their distributions might tend to be odd. Therefore, temporal drift is a complicating factor for forming document sub-collections. In this context, four different possible orderings were identified [2] for the documents to be added to sub-collections, based on their time stamps:

- **Chronological** ordering (**ch**)—the values of time stamps are increasing. The oldest documents are taken first. The newest documents are taken last. Using this order might make sense if the documents that passed the test of time prevail in the collection and the domain is well established, so it does not change significantly through time.
- **Reversed chronological** ordering (**rch**)—the values of time stamps are decreasing. The newest documents are taken first. The oldest documents are taken last. Using this order might make sense if the domain is being on the hype since recently and changes swiftly.

- **Bi-directional** ordering (**bd**)—the documents are taken from both sides of the temporal period of the time stamps, in turns. One first takes the oldest, then the latest, then the oldest from the remaining part, etc. Using this order might make sense if there is an estimated balance between the documents that passed the test of time and those, which became significant recently. A possible lowlight of going that way might be that the documents published in the middle of the time interval of a collection might not be picked.
- **Random** ordering (**rn**)—the documents are taken randomly. Using this order might make sense if there is no information about the terminological impacts of different documents and terminological change velocity in the domain.

In addition to the time of publication, one more aspect, that is important with regard to the significance and lifetime of a term, is the impact of a document, carrying this term, on domain description. Indeed, if a document influences the community and readership around the domain then the terms used in this document might be significant for the community. In scientific domains, where the documents are scholarly papers, an appropriate way to measure such an impact is the number of citations (c.f. Sect. 2.7). This number should however be balanced by the time passed since publication. Therefore, document citation frequency (e.g. the number of citations per year), might be an appropriate measure as it balances the position of a document on a time line with its impact on the domain description. Hence, one more potentially interesting ordering of documents for sub-collections is:

- **Descending citation frequency** ordering (**dcf**)—the documents are not taken in the order of their time stamps. One starts with the documents having the highest citation frequency[19] and continues down to the lowest citation frequency.

The research hypothesis we put forward and test in our experimental study is:

Hypothesis H4.1 (*dcf ordering*) The use of the **dcf** ordering is the **most balanced and effective** in coping with the aspects of compactness, performance, representativeness, stability, and sensitivity to accumulated regular noise (i–v), compared to all the orders based on time stamps.

We studied the influence of document ordering on:

- Terminological saturation; and
- Accumulated regular noise detection.

This influence was evaluated experimentally using the three document collections, DMKD, TIME, and DAC, described in Sect. 5.2.2.

Regarding the terminological saturation study, our specific experimental objectives were to find out if there was a particular ordering that resulted in balancing the aspects (i–iv) of compactness, performance, representativeness, and stability

[19] Citation frequency in our approach is computed as the number of citations during the years after publication divided by the number of years after publication minus one.

(Sect. 5.5.2). This study was performed using the DMKD, TIME, and DAC-Cleaned datasets.[20]

In the study of the sensitivity to accumulated regular noise (aspect (v)), our experimental objective was to find out if there was a particular ordering that was most sensitive to the appearance of such a noise in the data, hence detecting it quicker. This study was performed on the DAC collection datasets. In this experiment, the extracted bags of terms are not de-noised, which are further referred to as the DAC Naturelle datasets.

In the analysis of the experimental results, several measurable aspects were used to rank the orderings (Table 5.2). Ranking was done on a scale from 1 (the best) to 5 (the worst) for each combination of an individual aspect and document collection. These individual ranks were summed to determine the overall best ordering (Table 5.5) without giving preference (higher weight) to any individual aspect.

The experimental workflow and instrumental software toolset, outlined in Sect. 5.2.1, were used. Overall, the **Preparatory** phase was executed three times—one per document collection. The **Pre-processing** phase was executed 15 times—one per order of adding documents within a collection. The **Terms Extraction** and **Post-processing** phases were executed 20 times—15 times for different orderings and, additionally, five times for DAC Cleaned, after removing the accumulated regular noise.

5.5.2 Results of Terminological Saturation Study

The measurements at saturation points for all the five evaluated orderings (Sect. 5.5.1) are provided in Table 5.3 for comparison.

Individual Term Significance Thresholds (*Seps*). Individual term significance thresholds (*eps*) at the points of the detection of terminological saturation were compared in the context of this measurable aspect. The results are pictured in Fig. 5.11a–c.

As it could be seen in Fig. 5.11a, the highest *eps* values were achieved at saturation points if the **dcf** or **rch** orderings were used for DMKD collection. The saturation point for **rch** (D_7, D_6) was however achieved later than that for **dcf** (D_4, D_3). The use of **rn** was ranked in the middle of the scale—3. The **ch** and **bd** orderings were ranked the lowest as yielding the lowest *eps* values at the respective saturation points.

The ranking of the orderings regarding the TIME collection (Fig. 5.11b) was different. The highest rank was given to the **bd** ordering, though its saturation point was the latest. The second ranked was **rch**, the third ranked was **ch**, while **dcf** and **rn** were ranked the lowest (5).

[20] In the experiment using DAC datasets, the extracted bags of terms were de-noised at post-processing, using the **ARNR** algorithm (Sect. 4.7), which was further labelled as the DAC Cleaned datasets.

Table 5.3 The comparison of terminological saturation measurements for all orderings at corresponding saturation points

Ordering	Dataset pair (D_i, D_{i-1})	NET (B_i)	*eps* (B_i)	NRT (T_i) (*C*-Value > *eps*)	PRT, %	*thd* (T_i, T_{i-1})	ThdV
DMKD datasets							
ch	D_4, D_3	89,617	17.0000	3242	3.6176	16.2008	−0.8587
rn	D_5, D_4	124,331	20.5000	3739	3.0073	17.9221	−5.9448
rch	D_7, D_6	182,768	33.2842	3774	2.0649	18.9340	−18.0829
bd	D_{11}, D_{10}	319,441	24.0000	8054	2.5213	13.9341	−31.7709
dcf	D_4, D_3	109,076	19.6515	3399	3.1162	18.0366	−5.3458
TIME datasets							
ch	D_{12}, D_{11}	287,887	23.7744	7110	2.4697	18.1096	−6.0550
rn	D_{12}, D_{11}	252,767	22.0000	7129	2.8204	10.4441	−15.4888
rch	D_8, D_7	155,715	22.0000	4550	2.9220	20.1549	−29.9347
bd	D_{15}, D_{14}	329,240	24.0000	8723	2.6494	11.4488	−15.8609
dcf	D_9, D_8	174,103	19.0000	5493	2.5692	15.4377	−4.1768
DAC-cleaned datasets							
ch	D_{24}, D_{23}	514,364	15.5098	20,558	3.9968	13.6261	−2.2046
rn	D_{22}, D_{21}	481,885	15.5098	18,516	3.8424	10.8970	−6.8867
rch	D_{22}, D_{21}	486,649	15.5098	18,865	3.8765	10.9433	−7.6712
bd	D_{26}, D_{25}	551,165	15.5098	21,468	3.8950	3.7072	−12.8841
dcf	D_{25}, D_{24}	518,765	15.5098	19,991	3.8536	3.1238	−8.0866

In the DAC-Cleaned datasets experiment (Fig. 5.11c), saturation was reached quite lately with all orderings having the same *eps* values at their saturation points. Hence, all the orderings were ranked equally (1).

The summary ranking regarding this measurable aspect for the three collections was 4—**rch**; 7—**bd, dcf**; 9—**ch, rn**.

The Points of Entry into Saturation Zones (*SthdSP*). The rationale related to this measurable aspect was to find out which document ordering enabled reaching terminological saturation faster. Saturation was detected as explained by condition (3.4) in Sect. 3.1. Graphically, this condition meant that the curve of *thd* went below the curve of *eps* at the measurement point (D_i, D_{i-1})—which was the saturation point—and it never went up beyond the curve of *eps*, at least until the end of measurements.

For this measurable aspect, the *i*-s for different orderings were compared. In the tied cases, the *thd* values were additionally compared. The lower *thd* value in a tied case got better ranking. The results are pictured in Fig. 5.12a–c.

In the case of DMKD, Fig. 5.12a, one could have observed a tie between the **ch** and **dcf** orderings at (D_4, D_3). The **ch** ordering was ranked as the best as it arrived with a lower *thd* value at this point of measurement. The **dcf** ordering was ranked as the second best. The **rn** ordering was ranked as the third best at (D_5, D_4). The **rch**

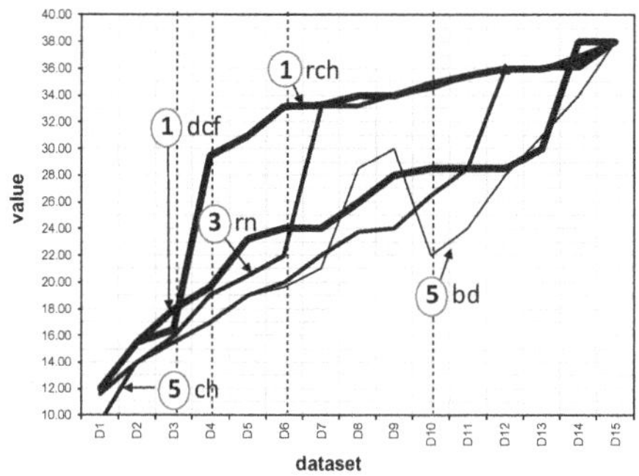

(a) DMKD datasets

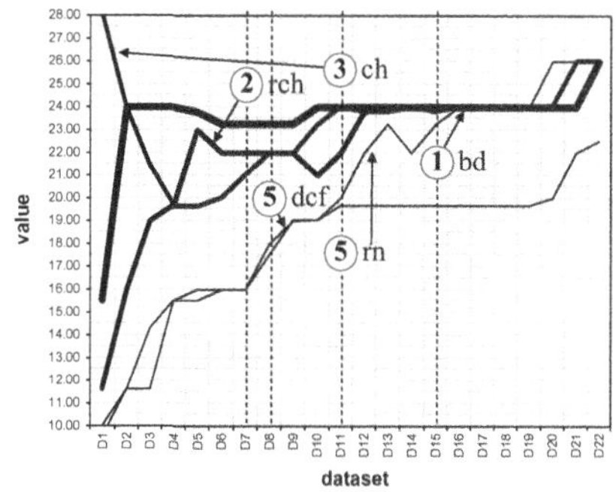

(b) TIME datasets

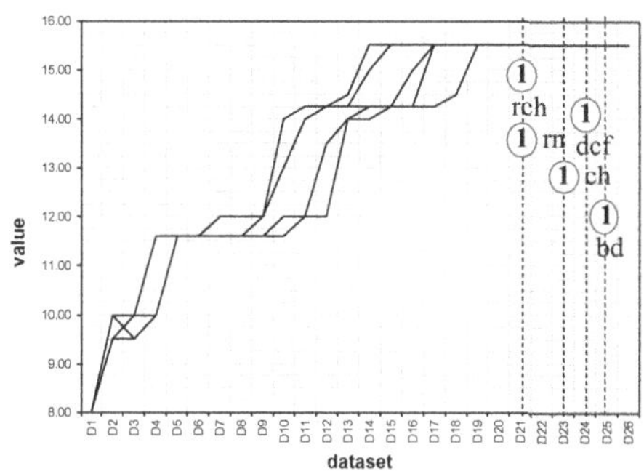

(c) DAC Cleaned datasets

Legend: The orderings are labelled by the legends linked to the curves by the arrows. The legends include also the ranks of the respective orderings given in circles. The charts render *eps* curves using different thicknesses. The thicker the line, the higher ranked is the ordering.

Fig. 5.11 Individual term significance thresholds (*eps*) for different document orderings

ordering was ranked as the fourth at (D_7, D_6). The **bd** ordering was ranked as the fifth at (D_{11}, D_{10}).

In the case of TIME, Fig. 5.12b, the rankings were as follows. The **rch** ordering was ranked as the first best at (D_8, D_7). The **dcf** ordering was ranked as the second best at (D_9, D_8). The tie at at (D_{12}, D_{11}) was resolved in favour of **rn** (third) that had the lower *thd* value compared to **ch** (fourth). Finally, the **bd** ordering was ranked as the fifth at (D_{16}, D_{15}).

In the case of DAC Cleaned, Fig. 5.12c, the rankings were the following. The tie at (D_{22}, D_{21}) was resolved in favour of **rn** (first) that had the lower *thd* value (10.8970, Table 5.3) than **rch** (second) with a slightly higher *thd* value (10. 9433). The **ch** ordering was ranked as the third at (D_{24}, D_{23}). The **dcf** ordering was ranked as the fourth at (D_{25}, D_{24}). The **bd** ordering was ranked as the fifth at (D_{26}, D_{25}).

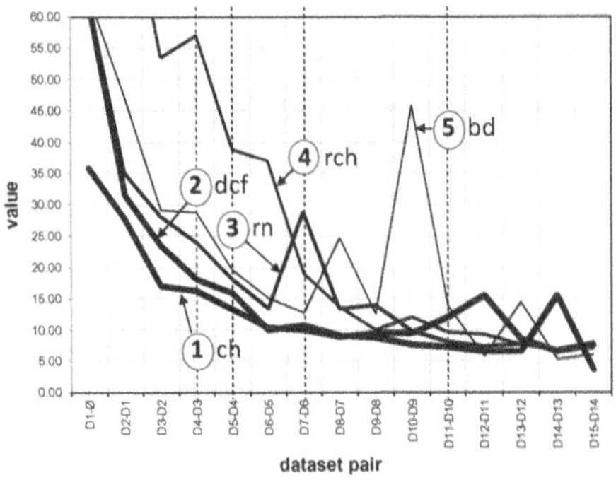

(a) DMKD datasets

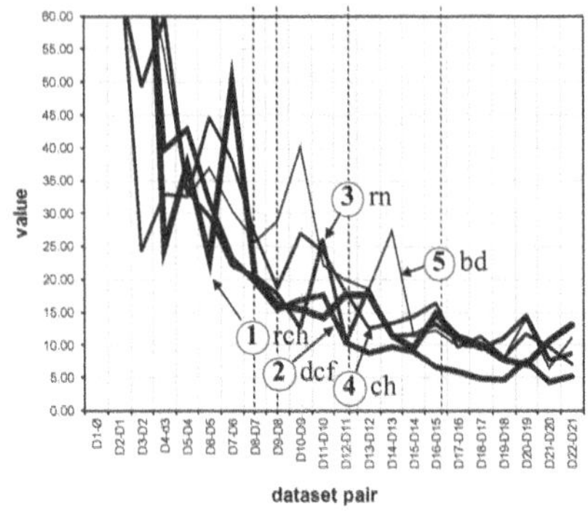

(b) TIME datasets

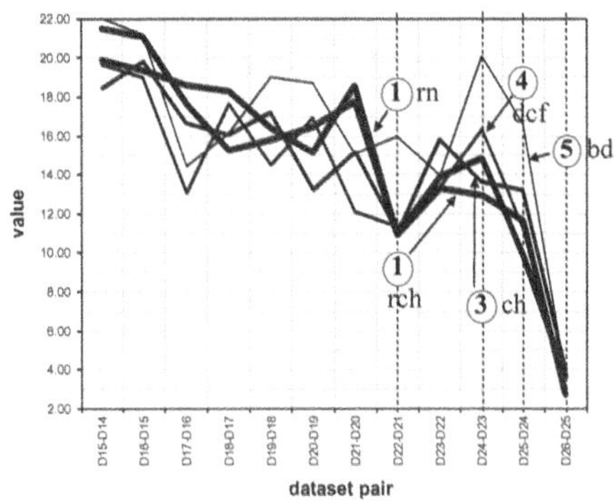

(c) DAC Cleaned datasets

Legend: The saturation points are labelled by vertical dashed lines. The orderings are labelled by the legends linked to the curves by the arrows. The legends include also the ranks of the respective orderings given in circles. These circles are placed on the saturation point lines to picture which ordering has what saturation point. The charts render ***thd*** curves using different thicknesses. The thicker the line, the higher ranked is the ordering.

Fig. 5.12 Terminological differences (*thd*) for different document orderings

The summary ranking regarding this measurable aspect for the three collections was 6—**ch**, **rch**; 7—**rn**; 8—**dcf**; 15—**bd**.

Average Proportions of Retained to all Extracted Terms (*SPrtSP*). The rationale related to this measurable aspect was to find out which document orderings resulted in yielding better representative bags of retained significant terms. The proportions were computed for all the iterations in the range of measurements and then averaged. The results are pictured in Fig. 5.13a–c.

In the case of DMKD, Fig. 5.13a, **ch** resulted in the highest proportion, which indicated the best representativeness of the respectively retained term-sets. The least representative result was obtained using **rch**. In the case of TIME, Fig. 5.13b, the result was substantially different, with **dcf** ranked as the highest and **ch**—as the

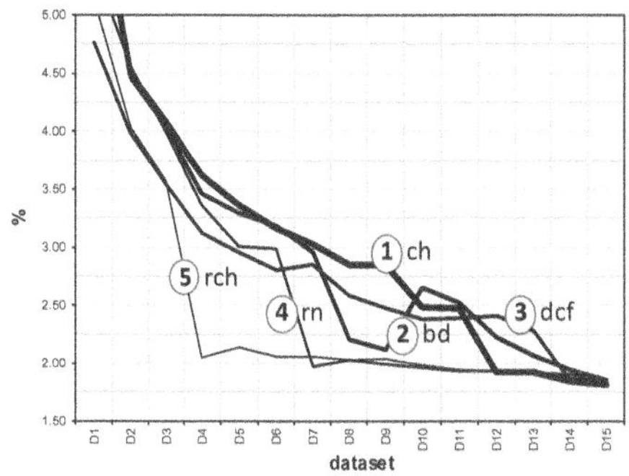

(a) DMKD datasets

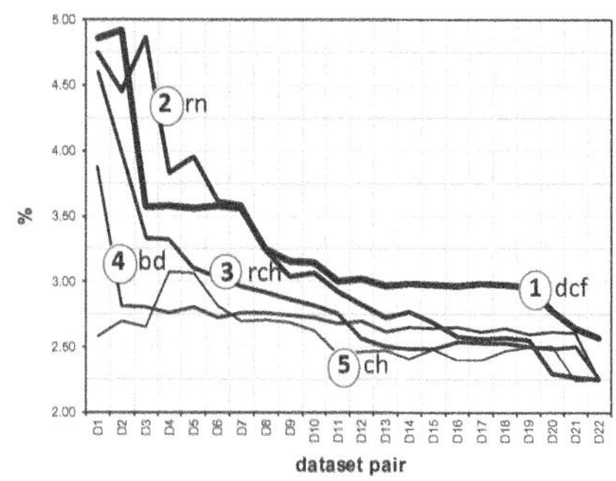

(b) TIME datasets

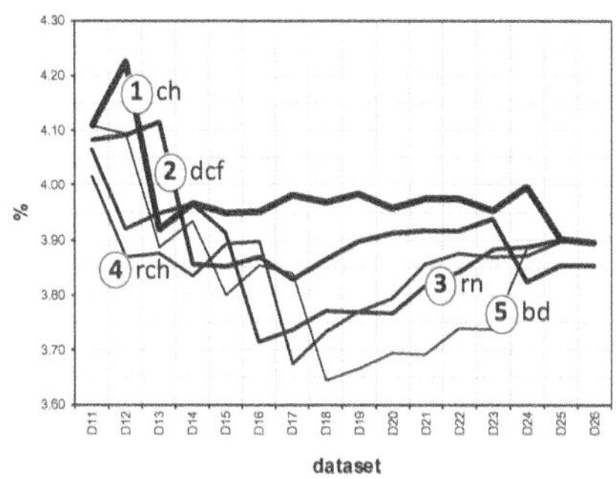

(c) DAC Cleaned datasets

Legend: The orderings are labelled by the legends placed on the corresponding curves. The curves represent all the pointwise proportion measurements per ordering. The ranks given in the circles are based on the averaged proportions. The charts render the curves using different thicknesses. The thicker the line, the higher ranked is the ordering.

Fig. 5.13 The average proportions of retained to all extracted terms for different document orderings

lowest. In the case of DAC Cleaned, Fig. 5.13c, **ch** was again ranked as the highest. Similar instabilities were observed for all the orders, but **dcf**, which was finally ranked the highest in the sum for all the three collections.

The summary ranking regarding this measurable aspect for the three collections was 6—**dcf**; 7—**ch**, 9—**rn**; 12—**rch**; 13—**bd**.

Integral Volatility of thd in Saturation Zones (*SthdV*). The rationale related to this measurable aspect was to find out which document ordering resulted in a more stable saturation process. It was supposed that the smoother the volatility curve was the least probable was the chance that *thd* went beyond *eps* in the measurements after the saturation point (within the saturation zone). The pointwise volatility (ThdV) values were computed as explained in Sect. 5.2.3. The absolute values were further summed for all the measurements in the saturation zones for different orders. The results are pictured in Fig. 5.14a–c.

The summary ranking regarding this measurable aspect for the three collections was 4—**ch**; 6—**dcf**; 9—**rn**; 11—**rch**; 15—**bd**.

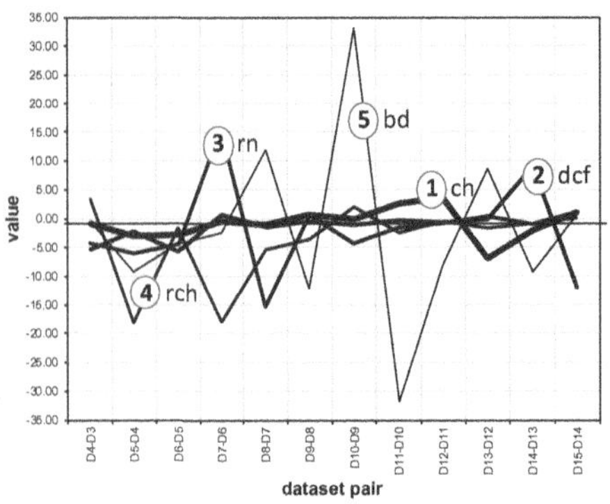

(a) DMKD datasets

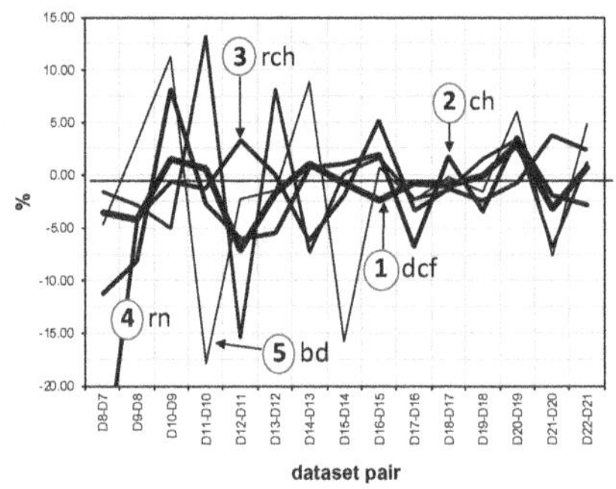

(b) TIME datasets

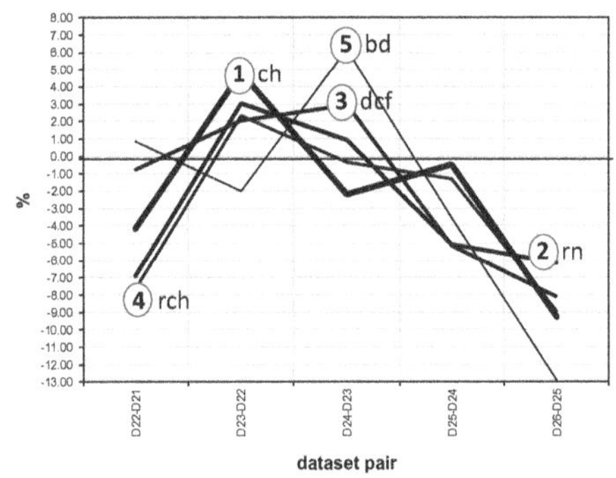

(c) DAC Cleaned datasets

Legend: The orderings are labelled by the legends placed on the curves or linked to the curves by the arrows. The curves represent all the pointwise volatility measurements per ordering. The legends include also the ranks of the respective orderings given in circles. These ranks are based on the sums of the pointwise volatility values in the respective saturation zones. The least integrally volatile ordering is ranked the highest. The charts render volatility curves using different thicknesses. The thicker the line, the higher ranked is the ordering.

Fig. 5.14 Integral volatilities of *thd* in saturation zones for different document orderings

5.5.3 Results of the Regular Noise Sensitivity Study

In this experiment, our focus was put on finding out which of the orderings resulted in higher sensitivity to accumulated regular noise. For that, it was evaluated how sharply and early the peaks of *eps* (*NepsPH*, *NepsPP*) and *thd* (*NthdPP*), indicating noise, occurred in the measurements using the DAC Naturelle datasets, which were substantially noisy. These measurements at the points of peak appearance are presented in Table 5.4.

An example of the observed accumulation of regular noise in this experiment is pictured in Fig. 5.15. This figure illustrates that the peak of *thd* is a more precise indicator of potential accumulation of regular noise as it occurs exactly at the measurement point where this accumulation becomes noticeable (Fig. 5.15a and b). Therefore, the amplitude of the *thd* peak is not that big compared to the amplitude of *eps* peak observed several measurement points later. The peak of *eps* points at, so to say, a mature bulk of regular noise that prevails in or even fully occupies the set of retained significant terms. This observation is supported by: (i) a much bigger amplitude of the peak, compared to *thd*; and (ii) a very substantial drop in the numbers of retained significant terms.

Figure 5.16 pictures our results in assessing and ranking the sensitivity of different orderings to the accumulation of regular noise. It was assessed using the measurable aspects presented in Sect. 5.2.3 (Table 5.2) based on the measurements of *eps* and *thd* for all the orders.

Table 5.4 DAC Naturelle

Ordering	Dataset (s)	NET (B_i)	$eps\ (B_i)$	NRT (T_i) (C-Value > eps)	PRT, %	$thd\ (T_i, T_{i-1})$	ThdV
(a) *eps peaks*							
ch	D_{11}	230,726	18,294.0372	34	0.0147	1.6057	−7.3833
rn	D_{12}	286,161	19,616.9148	36	0.0126	0.5045	−51.0163
rch	D_{16}	393,997	3107.6393	51	0.0129	3.3400	−0.3046
bd	D_{21}	463,660	352.0000	291	0.0628	18.4497	10.0066
dcf	D_{11}	218,764	19,338.4980	33	0.0151	1.4475	−5.7119
(b) *thd peaks*							
ch	D_5, D_4	277,775	32.0000	3191	1.1488	154.2663	82.8619
rn	D_{11}, D_{10}	265,540	48.0000	1934	0.7283	51.5208	87.8267
rch	D_{11}, D_{10}	277,775	32.0000	3191	1.1488	107.9931	86.4081
bd	D_{10}, D_9	234,634	23.7744	4378	1.8659	81.0028	62.2701
dcf	D_5, D_4	106,814	24.0000	1906	1.7844	149.3096	83.8976

The comparison of noise indications for all the orderings

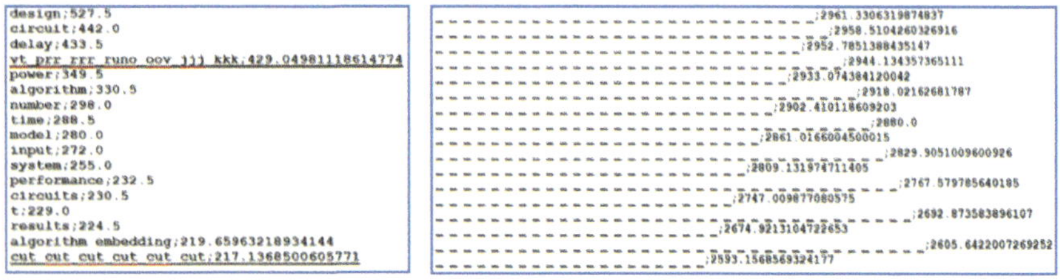

(a) T_4 extracted from D_4, $thd = 65.42$ (b) T_5 extracted from D_5, $thd = 149.31$

Legend: (a) the top part of T_4 in which regular noise, though present (underscored), has not yet been accumulated; (b) the top part of T_5 in which regular noise has been accumulated. At (D_5, D_4) the peak of *thd* is observed – the increase from 65.42 to 149.31.

Fig. 5.15 An example of the accumulation of regular noise in the sets of retained significant terms extracted from DAC Naturelle using the **dcf** ordering

With respect to the amplitude of the *eps* peak (Fig. 5.16a), the most sensitive ordering was **rn**, ranked first. It was followed by the **dcf** (second) and **ch** (third) orderings. All these three orderings demonstrated the peaks of relatively similar height, compared to the much worse sensitivity of **rch** (fourth) and **bd** (fifth).

As for the measurement points of the *eps* peaks (Fig. 5.16b), the earliest, hence the most sensitive, were the **dcf** and **ch** orderings that both peaked at D_{11}. Therefore, both were ranked first. The **rn** ordering peaked at D_{12}, ranked third. The **rch** ordering peaked at D_{16}, ranked fourth. The **bd** ordering peaked at D_{21}, ranked fifth.

Sensitivity ranking based on the measurement points of the *thd* peaks (Fig. 5.16c) revealed the following. The most sensitive were **dcf** and **ch** that both peaked the earliest at (D_5, D_4) with very similar amplitudes. The third ranked was **bd** that peaked at (D_{10}, D_9). The fifth ranked were **rch** and **rn** that peaked at (D_{11}, D_{10}).

The summary ranking with respect to the sensitivity to accumulated regular noise by the three measurable aspects was 4—**dcf**; 5—**ch**; 9—**rn**; 13—**rch**, **bd**.

5.5.4 Overall Ranking and Recommendation

This Section summarizes and compares our findings and rankings after analysing the results of the cross-evaluation experiments of different document orderings for: (i) measuring terminological saturation; and (ii) detecting accumulated regular noise. The summary (Table 5.5) is structured along the document collections and measurable aspects used in these experiments.

In the orderings columns, the ranks per aspect are given for the respective individual orderings. The ranks are summed per collection and the ranks of the orderings are assigned per collection using the same scale: 1—the highest to 5—the lowest. Finally, the individual ranks (per measurable aspects) and collection ranks are summed at the bottom of the table. Hence, the sum of the individual ranks indicates how good, integrally, was an ordering with respect to the specified measurable

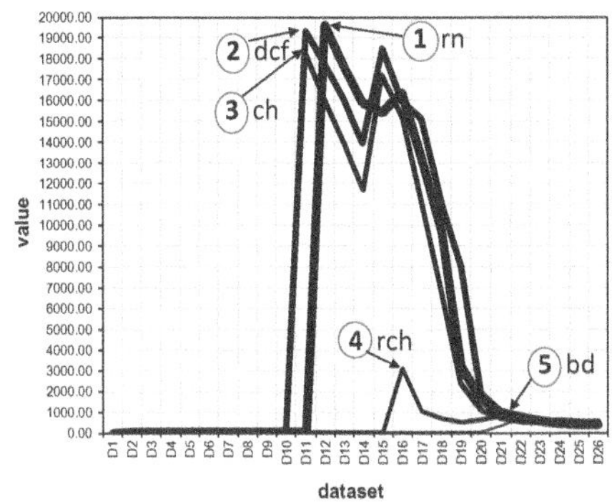

(a) Height of *eps* peak (*NepsPH*)

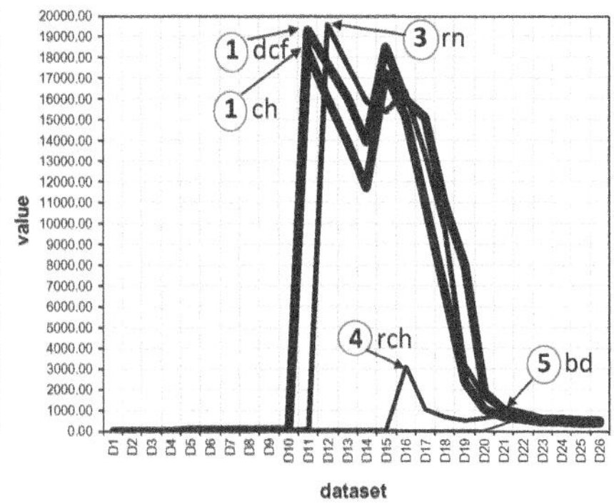

(b) Point of *eps* peak (*NepsPP*)

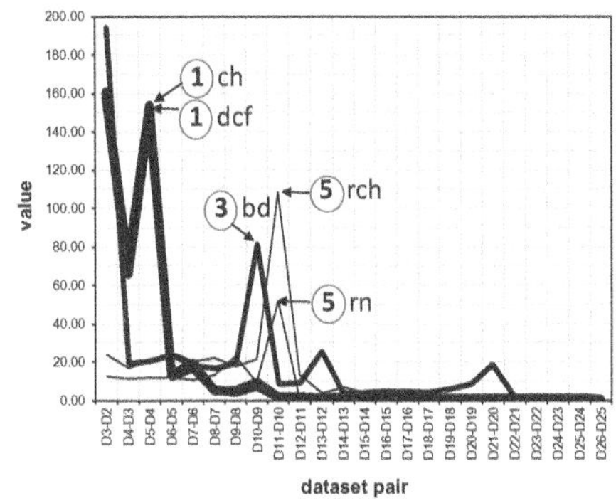

(c) Point of *thd* peak (*NthdPP*)

Legend: The orderings are labelled by the legends linked to the curves by the arrows. The curves represent all the pointwise *eps* (a) and (b), and *thd* (c) measurements per ordering. The legends include also the ranks of the respective orderings given in circles. These ranks are assigned based on the ranking rules of the measurable criteria. The charts render sensitivity curves using different thicknesses. The thicker the line, the higher ranked is the ordering.

Fig. 5.16 Sensitivity to accumulated regular noise using different document orderings

aspects. The sum of the collection-based ranks outlines how balanced was ordering across all the collections and experiments.

Judging by individual aspect ranks, **dcf** demonstrated the **best overall capacity to balance** all the required evaluation aspects (i–v, Sect. 5.5.1) with the lowest sum of individual ranks—31. The second best was **ch**—33. However, regarding the cross-collection balance, **dcf** and **ch** appeared to be equally the **highest ranked** in all the experiments with equal sums of collection-based ranks—7. A more detailed comparison of **ch** to **dcf** revealed that: (i) **dcf** (rank = 4) lost to **ch** (rank = 1) in the experiment with the noisiest DAC collection; however, (ii) **dcf** appeared to be the most sensitive to accumulated regular noise (rank = 1), with **ch** been the follower (rank = 2). Hence, **dcf** was not very good on DAC Cleaned due to yet

Table 5.5 The summary of the cross-evaluation of different document orderings based on the specified measurable aspects

Collection		Ordering				
Measurable aspect	Ranking rule	ch	rch	rn	bd	dcf
Influence on terminological saturation						
DMKD						
Seps	The higher–the better	5	1	3	5	1
SthdSP	The earlier–the better	1	4	3	5	2
SPrtSP	The higher–the better	1	5	4	2	3
SthdV	The lower–the better	1	4	3	5	2
Sum		8	14	13	17	8
Collection rank		1	4	3	5	1
TIME						
Seps	The higher–the better	3	2	5	1	5
SthdSP	The earlier–the better	4	1	3	5	2
SPrtSP	The higher–the better	5	3	2	4	1
SthdV	The lower–the better	2	3	4	5	1
Sum		14	9	14	15	9
Collection rank		3	1	3	5	1
DAC Cleaned						
Seps	The higher–the better	1	1	1	1	1
SthdSP	The earlier–the better	3	1	1	5	4
SPrtSP	The higher–the better	1	4	3	5	2
SthdV	The lower–the better	1	4	2	5	3
Sum		6	10	7	16	10
Collection rank		1	4	2	5	4
Sum of measurable aspect ranks (saturation)		28	33	34	48	27
Sum of collection ranks		5	9	8	15	6
Sensitivity to accumulated regular noise						
DAC Naturelle						
NepsPH	The higher–the better	3	4	1	5	2
NepsPP	The earlier–the better	1	4	3	5	1
NthdPP	The earlier–the better	1	5	5	3	1
Sum of measurable aspect ranks (noise)		5	13	9	13	4
Collection rank		2	5	3	5	1
Total sum of measurable aspect ranks		33	46	43	61	31
Total sum of collection ranks		7	14	11	20	7
Final rank		2	4	3	5	1

remaining excessive noise that was on the other hand best detected by **dcf**. Based on this analysis, **dcf** is preferred to **ch** in the final total ranking.

The rest of the orderings appeared to be the outliers as revealed by all total rankings in Table 5.5.

The ranking based on the specified measurable aspects and results of our cross-evaluation experiments implies that the **dcf** ordering is recommended as the best balancing the requirements of compactness, performance, representativeness, stability, and sensitivity to accumulated regular noise.

5.6 Influence of Term Grouping

In this section, our experiments on grouping terms[21] are reported and discussed. The focus in these experiments was put on the study of the influence of terms grouping on terminological saturation. The specifics of the experimental setup is presented in Sect. 5.6.1. The results of the experiments are discussed in Sect. 5.6.2. The ranking of these results and offered recommendations conclude the discussion in Sect. 5.6.3.

5.6.1 Particularities in Experimental Settings

The refined **`R-THD`** algorithm (Sect. 4.6.4) for measuring terminological difference with an account for terms grouping against the baseline **`THD`** algorithm (Sect. 4.5) was evaluated. Experiments were conducted using the workflow reported in Sect. 5.5.1. The workflow was applied to the three real document collections, TIME, DMKD, and DAC-cleaned, described in Sect. 5.5.2.

The objective of our experiments was to verify if using the **`R-THD`** algorithm, compared to the use of the baseline **`THD`** algorithm, resulted in less iterations for detecting terminological saturation and yielded less volatile values of terminological differences between the bags of retained significant terms in iterations. One more objective was to find out which string similarity measure (SSM) is the most productive in the context of detecting terminological saturation with terms grouping.

The datasets, generated from the real document collections, mentioned above, were fed into the (i) refined **`R-THD`** algorithm and (ii) baseline **`THD`** algorithm. For every pair of the bags of term candidates (B_i, B_{i-1}) extracted from document collections, the algorithms were applied:

- **`R-THD`**—16 times, one per individual SSM M^{22} (Sect. 4.6.1) and per individual term similarity threshold value *th* (Sect. 4.6.2, Table 4.5); and

[21] The material of this section is based on our publication [8].

[22] The functions for all the four selected SSMs (Sect. 4.8) return real values within [0, 1].

- **THD**—one time, as the baseline algorithm did not use string similarity measures and term similarity thresholds.

The values of the: (i) number of retained significant terms (NRT); (ii) absolute terminological difference (*thd*); and (iii) time spent by the **STG** algorithm (sec) to group similar terms were measured.

To verify the correctness of our SSM, **STG**, and **R-THD** implementations, it was checked if the results returned by **R-THD** were sufficiently similar to that produced by the baseline algorithm (**THD**) when *th* equalled to 1.00. This threshold value forced our SSM to regard only identical strings as similar terms. Hence, the results returned by **R-THD** should ideally have been the same as the results by **THD**.

5.6.2 Results and Discussion

The measurements taken in the experiments on different collections and with different th values are visualized below. The full details of these measurements are available publicly online.[23]

The thd curves are pictured in Fig. 5.17 for TIME, Fig. 5.18 for DMKD, and 5.19 for DAC Cleaned. These curves testify that **R-THD** spent less iterations than the baseline (**THD**) for detecting terminological saturation. This observation was valid for any tested SMM (*M*) or similarity threshold (*th*).

It could also be observed that, overall, all the evaluated SSM facilitated adequately retaining significant terms by **R-THD** from the datasets generated from all three document collections. The co-locations of terminology contribution peaks at the diagrams (a)–(d) in Figs. 5.17, 5.18 and 5.19 support this conclusion. In particular, in the case of the Max *th* values (Figs. 5.17, 5.18 and 5.19d), all the saturation curves built using the SSM follow the shape of the baseline **THD** curve closely. Specifically, the SSM curves demonstrate terminology contribution peaks exactly at the same iterations as the baseline does. For all the three collections, the Sørensen-Dice measure was the most sensitive, as it facilitated the production of the peaks with the biggest amplitude, excluding the baseline.

Figures 5.20, 5.21 and 5.22 picture the times spent by the **STG** algorithm for finding and grouping similar terms for different *th*.

As it is clear from Figs. 5.20, 5.21 and 5.22, the use of SSM and adding **STG** to the pipeline for terminological saturation measurement substantially increased the time spent for computation. The expenses of run time grew with the increase of *th*, depending also on the sizes of the processed bags of terms, from several hundreds (*Min* thresholds) to several thousands (*Max* thresholds) of seconds. A closer look at Figs. 5.20, 5.21 and 5.22 reveals that Jaccard and Sørensen-Dice were times less expensive than Jaro and Jaro-Winkler. Sørensen-Dice spent substantially more time

[23] https://github.com/OntoElect/Experiments/tree/master/STG. File names are {TIME, DMKD-300, DAC-cleaned}-Results-Alltogether-{min, ave, ave2, max, 1}-th.xlsx.

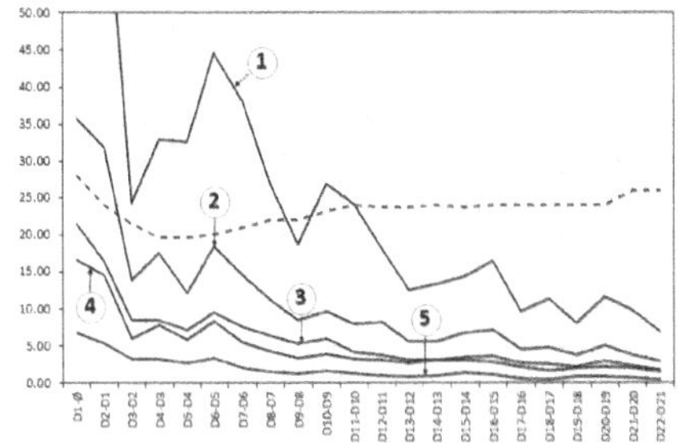

(a) *Min* term similarity thresholds

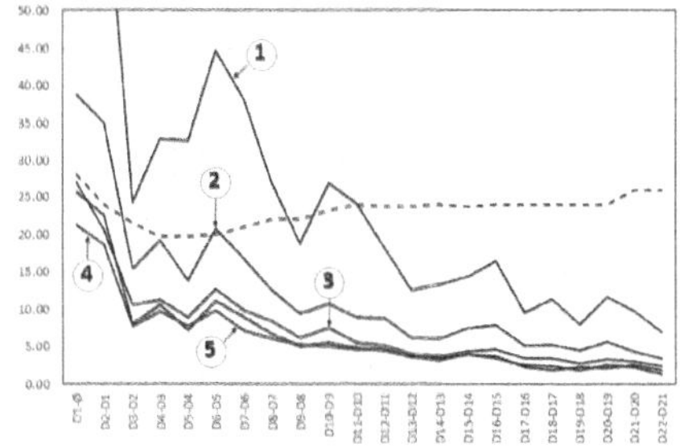

(b) *Ave*-1 term similarity thresholds

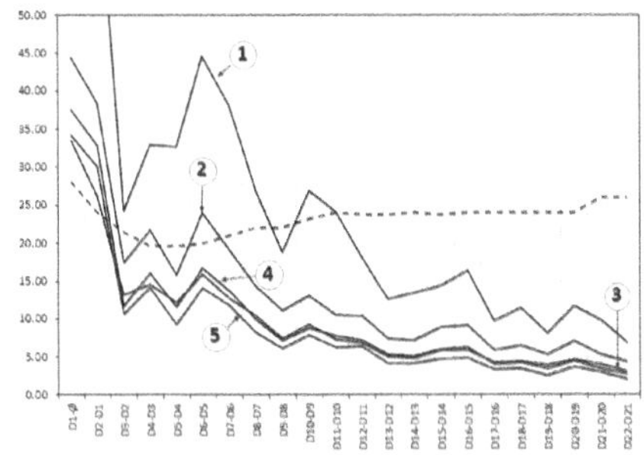

(c) *Ave*-2 term similarity thresholds

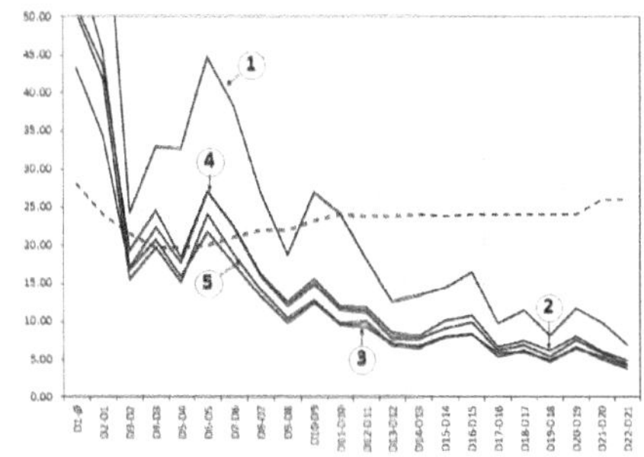

(d) *Max* terms similarity thresholds

Legend: 1 – Baseline; 2 - Sørensen-Dice; 3 – Jaccard; 4 – Jaro; 5 – Jaro-Winkler; dashed curve – *eps*

Fig. 5.17 Terminological saturation (*thd*) measurements for TIME using different *th* values

than Jaccard. However, Sørensen-Dice outperformed Jaccard in its sensitivity to terminology contribution peaks and the numbers of retained significant terms.

The proportions of the numbers of retained significant terms to all extracted term candidates at saturation detection points, computed for different *th* values, are shown in Fig. 5.23. The figure testifies that the second-highest proportions, after the baseline, across all the collections and *th* values were by Sørensen-Dice.

Finally, for testing the correctness of our SSM, **STG**, and **R-THD** implementations, the refined saturation measurement pipeline was cross-evaluated against the baseline using all three collections of documents. The measurements were done in a few iterations around the saturation points. It was checked if the refined pipeline:

- Detected terminological saturation at the same iteration as the baseline. Therefore, *eps* and *thd* values were assessed.
- Retained a similar number of significant terms as the baseline. Therefore, the NRT values were assessed.

Additionally, the times spent accomplishing **STG** were assessed and compared for different SMM.

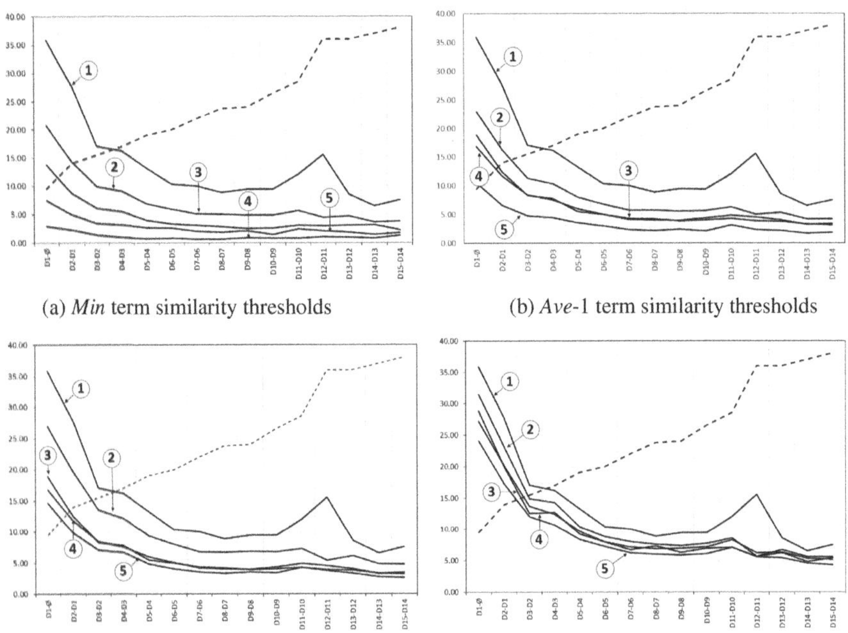

(a) *Min* term similarity thresholds

(b) *Ave*-1 term similarity thresholds

(c) *Ave*-2 term similarity thresholds

(d) *Max* term similarity thresholds

Legend: 1 – Baseline; 2 - Sørensen-Dice; 3 – Jaccard; 4 – Jaro; 5 – Jaro-Winkler; dashed curve – ***eps***

Fig. 5.18 Terminological saturation (*thd*) measurements for DMKD using different *th* values

The results of the experiment with the DMKD bags of terms are visualized in Fig. 5.24. The results for TIME and DAC-cleaned[24] were practically identical to DMKD case.

Figure 5.24a and b testifies that the pipelines using Jaro and Jaro-Winkler yielded the measurements that repeated the results of the baseline in *thd*, *eps*, and NRT values. Sørensen-Dice behaved similarly to Jaro and Jaro-Winkler before the point of detecting saturation. After this point, it returned insignificantly lower *thd* and NRT values. As the measurements after the saturation point were of marginal interest, it could be regarded that the results by Sørensen-Dice were similarly correct as those by Jaro and Jaro-Winkler. The pipeline that used Jaccard returned noticeably lower *thd* and NRT in all iterations. Furthermore, Jaccard detected saturation one iteration earlier than the baseline and the rest of the SSMs, which should not have happened for $th = 1.00$.

As pictured in Fig. 5.24c, Jaro and Jaro-Winkler pipelines were computationally ~ 3 times more expensive than Sørensen-Dice. In turn, Sørensen-Dice was ~ 3 times

[24] These results could be accessed at https://github.com/OntoElect/Experiments/tree/master/STG. File names are {TIME, DMKD-300, DAC-cleaned}-Results-Alltogether-1-th.xlsx.

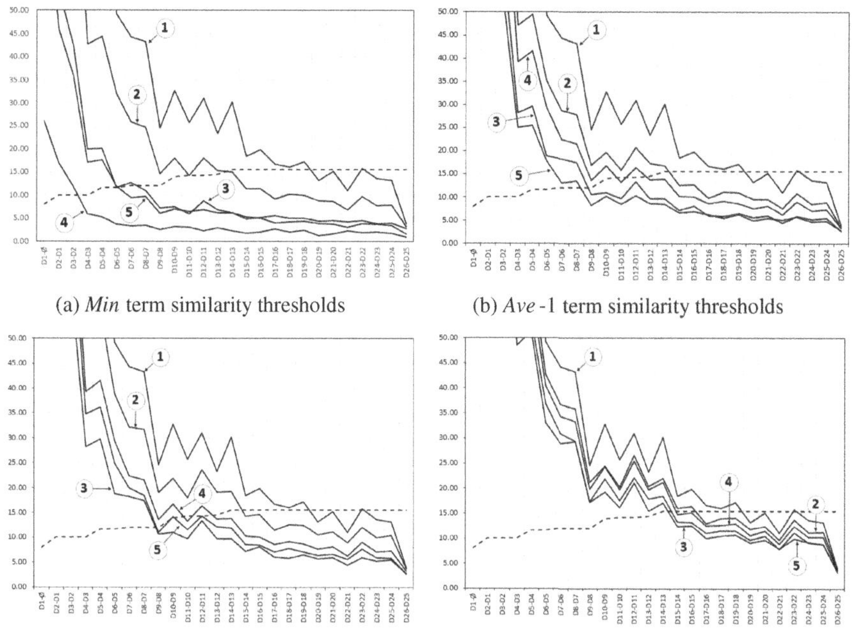

(a) *Min* term similarity thresholds
(b) *Ave*-1 term similarity thresholds
(c) *Ave*-2 term similarity thresholds
(d) *Max* term similarity thresholds

Legend: 1 – Baseline; 2 - Sørensen-Dice; 3 – Jaccard; 4 – Jaro; 5 – Jaro-Winkler; dashed curve – eps

Fig. 5.19 Terminological saturation (*thd*) measurements for DAC-Cleaned using different *th* values

more expensive than Jaccard. Hence, among the SSM that yielded results similar to the baseline, Sørensen-Dice demonstrated the best performance.

5.6.3 Overall Ranking and Recommendation

The results of our cross-evaluation experiments on terms grouping are summarized in Table 5.6 in the form of the rankings. The performance of all the SSM pipelines and the baseline was ranked using the scale from 1 (the best) to 5 (the worst) for every *th* point (*Min*, *Ave*1, *Ave*2, *Max*) with respect to every document collection. Average rankings were computed for all four *th* points.

The following aspects for ranking the SSM pipelines were evaluated: (i) the number of iterations spent for detecting terminological saturation, the less–the better (Figs. 5.17, 5.18 and 5.19); (ii) the number of retained significant terms, the more–he better (Fig. 5.23); and (iii) the computation time spent by an SSM pipeline, the less–the better (Figs. 5.20, 5.21 and 5.22).

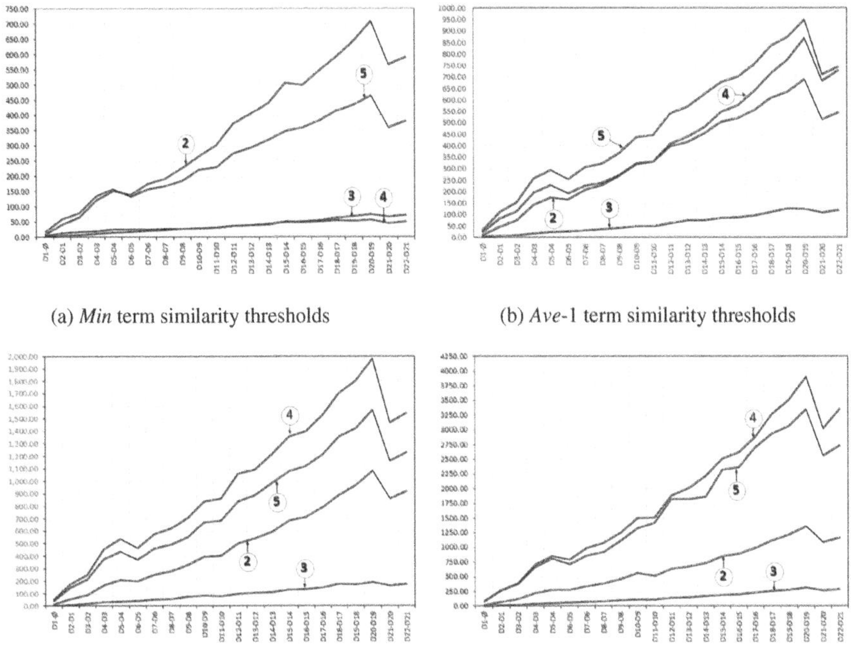

(a) *Min* term similarity thresholds (b) *Ave*-1 term similarity thresholds

(c) *Ave*-2 term similarity thresholds (d) *Max* term similarity thresholds

Legend: 2 - Sørensen-Dice; 3 – Jaccard; 4 – Jaro; 5 – Jaro-Winkler

Fig. 5.20 Time (s) spent by the **STG** algorithm for grouping similar terms in the TIME bags of terms

Table 5.7 presents the performance indices for the cross-evaluated SSM and the baseline with regard to the four *th* points and their average values across the used document collections. The indices for following two cases are provided: (a) execution time-based ranking was taken into account; and (b) execution time-based ranking was excluded. The latter case was provided for comparing the SSM if the minimization of computational overhead was of low priority. The values were calculated by summing all the ranks for different collections and aspects taken from the corresponding th point rows of Table 5.6. These sums were further subtracted from the highest rank value[25] in order to revert to a more natural "the higher–the better" scale in Table 5.7. Performance indices are also pictured in Fig. 5.25.

The evaluation of correctness at $th = 1.00$ revealed that the SSM pipeline, that demonstrated the balance of correctness before the saturation point and computational efficiency, was Sørensen-Dice. Jaro and Jaro-Winkler, though provided correct results (similar to the baseline), spent times more time to accomplish computations.

[25] The rank value is the sum of all ranks for a method within a particular threshold in Table 5.6. The highest rank value indicates the lowest performance. For (a) it equals to 12, which is for Sørensen-Dice at Min threshold. For (b) it equals to 4.

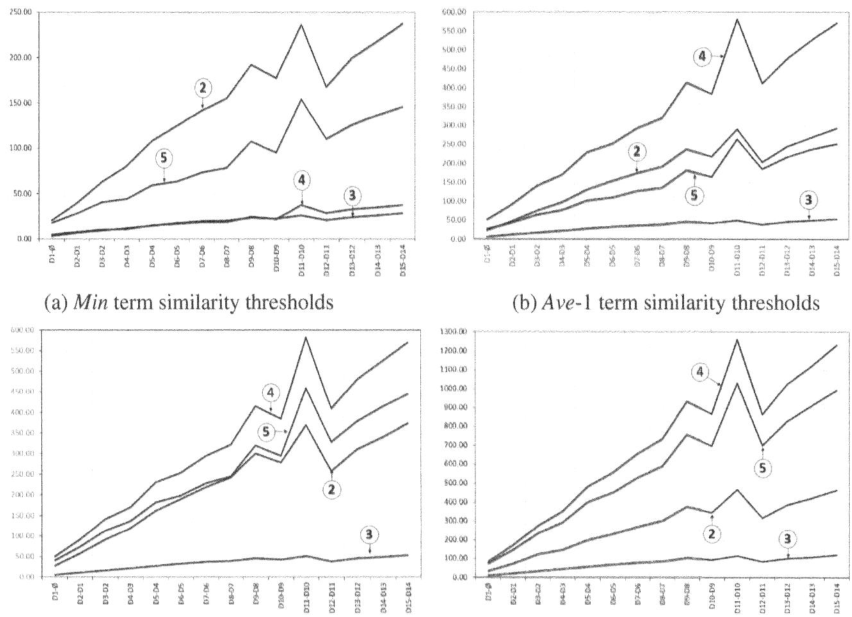

(a) *Min* term similarity thresholds

(b) *Ave*-1 term similarity thresholds

(c) *Ave*-2 term similarity thresholds

(d) *Max* term similarity thresholds

Legend: 2 - Sørensen-Dice; 3 – Jaccard; 4 – Jaro; 5 – Jaro-Winkler

Fig. 5.21 Time (s) spent by the **STG** algorithm for grouping similar terms in the DMKD bags of terms

The pipeline that used Jaccard did not return results similar to the baseline. Hence, it was incorrect.

It was also useful to compare the performance of the cross-evaluated SSM for the cases of (a) accounting and (b) not accounting for the time spent in computation—Table 5.7. The results of the comparison are visualized in Fig. 5.25.

For (a) the conclusions are the following. Jaccard, been the most lightweight SSM in terms of time for computation (Figs. 5.20, 5.21 and 5.22), has received the highest SSM performance index. Of course, it did not outperform the baseline as: (i) it took additional time for **STG**; and (ii) it retained less significant terms.[26] Jaccard was the best balanced on all evaluation aspects, compared to the other evaluated SSM. One important shortcoming of Jaccard was that it performed incorrectly at $th = 1.00$, as it returned results that noticeably differed from that of the baseline. Therefore, the use of Jaccard may be recommended when low *th* (like *Min* or *Ave*1) and hard constraints on the computation time are applied. Performance indices were also good for Sørensen-Dice and Jaro-Winkler. These pipelines also worked correctly at $th = 1.00$. Hence, these two SSM could be regarded as mutually complementary

[26] Which should be so as the baseline does not group terms. Therefore, any pipeline that groups similar terms retains less significant terms.

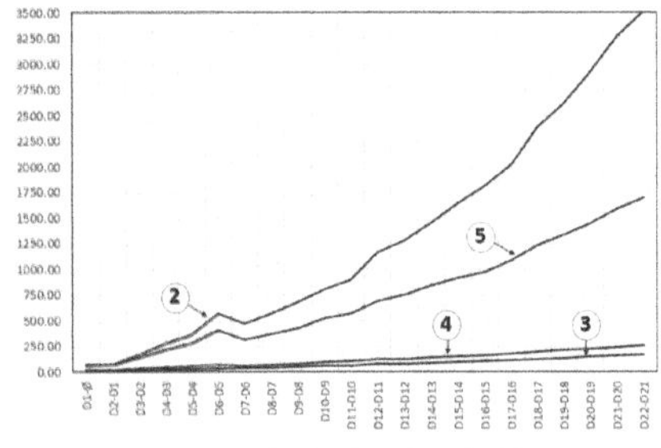

(a) *Min* term similarity thresholds

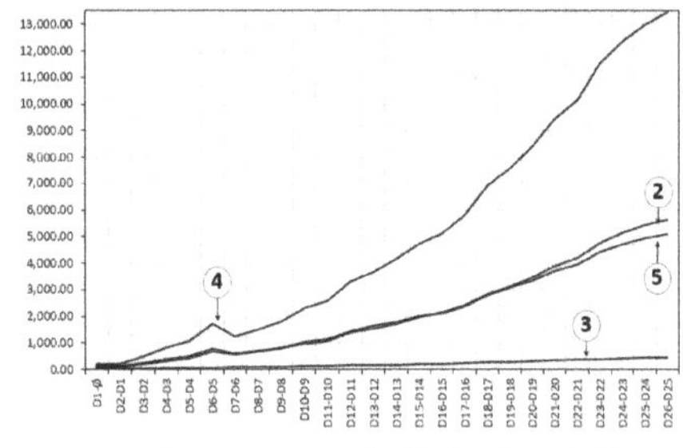

(b) *Ave*-1 term similarity thresholds

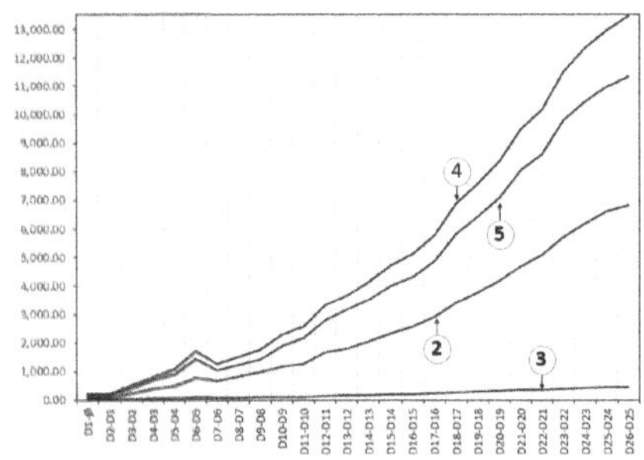

(c) *Ave*-2 term similarity thresholds

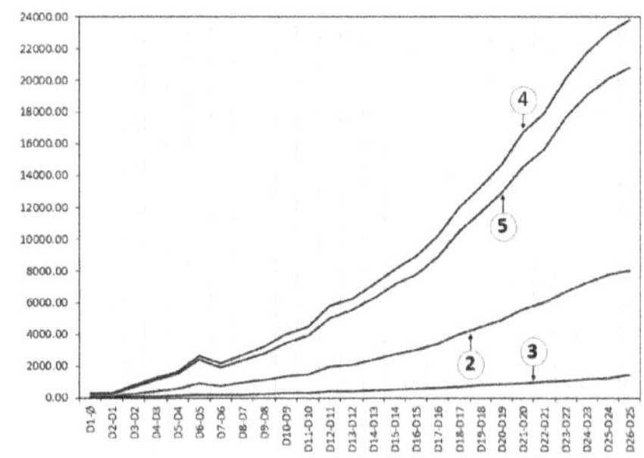

(d) *Max* term similarity thresholds

Legend: 2 - Sørensen-Dice; 3 – Jaccard; 4 – Jaro; 5 – Jaro-Winkler

Fig. 5.22 Time (s) spent by the **STG** algorithm for grouping similar terms in the DAC-cleaned bags of terms

because: (i) Jaro-Winkler outperformed Sørensen-Dice at lower *th*: *Min* or *Ave*1; (ii) Sørensen-Dice outperformed Jaro-Winkler at higher *th*: *Ave*2 or *Max*. Jaro was a negative outlier and is not recommended for use in case (a).

For (b) the conclusions are different from (a). It is clear from Fig. 5.25b that all the cross-evaluated SSM outperformed the baseline on average and for $th = Max$. Jaro-Winkler demonstrated the best performance for *th* equal to *Min* and *Ave*1 but gave up to Jaro for *th* equal to *Ave*2 and *Max*. Jaro-Winkler was also outperformed by Sørensen-Dice for *th* equal to *Max*. Nevertheless, Jaro-Winkler demonstrated the most balanced performance across all the four *th*, which is testified by the Avg value in Fig. 5.25b. Jaccard was regarded as a negative outlier and therefore could not be recommended for use.

If the conclusions for the cases (a) and (b) are combined, the following recommendation could be given. At the expense of a substantially higher execution time, the refined pipeline using **R-THD**, **STG**, and Jaro-Winkler SSM (for all *th* except *Max*) or using Sørensen-Dice (for $th = Max$) is our recommended configurations for measuring terminological saturation. Jaro-Winkler is the first choice because its performance is the most balanced across all the evaluated th points.

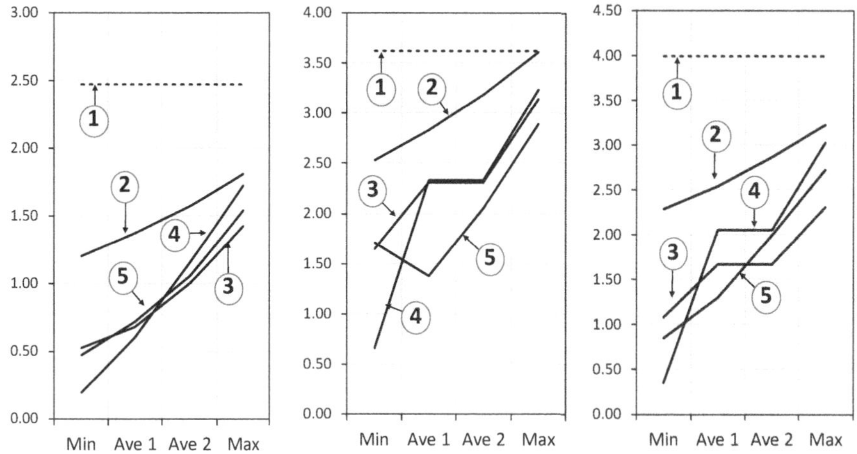

(a) TIME collection (b) DMKD-300 collection (c) DAC-cleaned collection

Legend: 1 – Baseline; 2 - Sørensen-Dice; 3 – Jaccard; 4 – Jaro; 5 – Jaro-Winkler

Fig. 5.23 The proportions (percent) of retained to all extracted terms for different term similarity thresholds per document collections

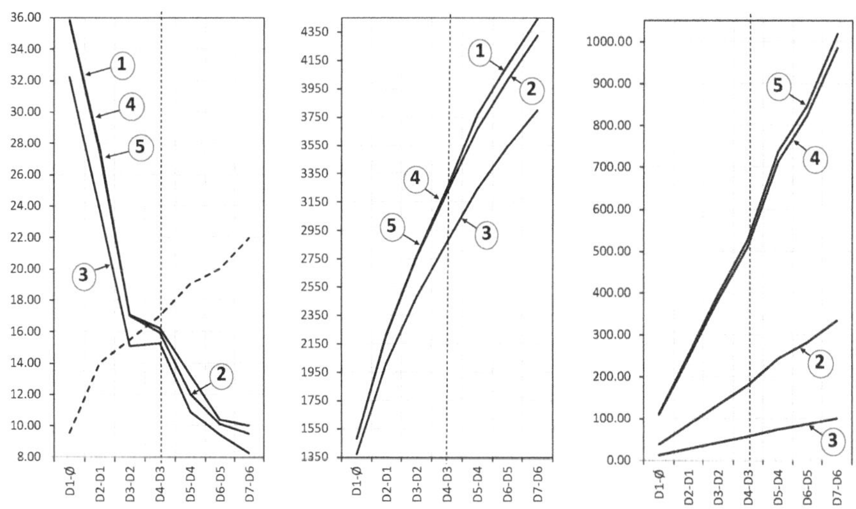

(a) *thd* values (b) numbers of retained significant terms (c) time taken by **STG**

Legend: 1 – Baseline; 2 – Sørensen-Dice; 3 – Jaccard; 4 – Jaro; 5 – Jaro-Winkler; dashed curve – *eps*; vertical dashed line – saturation point

Fig. 5.24 Evaluation of the refined (**R-THD**) pipelines with different SSM at $th = 1.00$ on the DMKD bags of terms

Table 5.6 The ranking of the evaluated SSMs

		Rank (1–5)				
Aspect	*th*	Baseline	Sørensen-Dice	Jaccard	Jaro	Jaro-Winkler
TIME collection						
Less iterations for detecting saturation	*Min*	5	4	1	1	1
	*Ave*1	5	4	1	1	1
	*Ave*2	5	4	1	1	1
	Max	5	3	1	3	1
	Average	5	3.75	1	1.5	1
More significant terms retained	*Min*	1	2	3	5	4
	*Ave*1	1	2	4	5	3
	*Ave*2	1	2	5	3	4
	Max	1	2	5	3	4
	Average	1	2	4.25	4	3.75
Less time spent	*Min*	1	5	3	2	4
	*Ave*1	1	4	2	3	5
	*Ave*2	1	3	2	5	4
	Max	1	3	2	5	4
	Average	1	3.75	2.25	3.75	4.25
DMKD collection						
Less iterations for detecting saturation	*Min*	5	4	3	1	1
	*Ave*1	5	4	1	1	1
	*Ave*2	5	4	1	1	1
	Max	5	1	1	1	1
	Average	5	3.25	1.5	1	1
More significant terms retained	*Min*	1	2	3	5	4
	*Ave*1	1	2	4	5	3
	*Ave*2	1	2	5	3	4
	Max	1	2	5	3	4
	Average	1	2	4.25	4	3.75
Less time spent	*Min*	1	5	2	3	4
	*Ave*1	1	4	2	5	3
	*Ave*2	1	3	2	5	4
	Max	1	3	2	5	4
	Average	1	3.75	2	4.5	3.75
DAC-cleaned collection						
Less iterations for detecting saturation	*Min*	5	4	3	1	2
	*Ave*1	5	4	1	3	1

(continued)

Table 5.6 (continued)

		Rank (1–5)				
Aspect	*th*	Baseline	Sørensen-Dice	Jaccard	Jaro	Jaro-Winkler
	Ave2	5	4	1	3	1
	Max	5	4	1	1	1
	Average	5	4	1.5	2	1.25
More significant terms retained	*Min*	1	2	3	5	4
	Ave1	1	2	4	3	5
	Ave2	1	2	5	3	4
	Max	1	2	5	3	4
	Average	1	2	4.25	3.5	4.25
Less time spent	*Min*	1	5	2	3	4
	Ave1	1	4	2	5	3
	Ave2	1	3	2	5	4
	Max	1	3	2	5	4
	Average	1	3.75	2	4.5	3.75

Table 5.7 The performance indices of the evaluated SSM with and without account for execution time

th	Baseline THD	Sørensen-Dice	Jaccard	Jaro	Jaro-Winkler
(a) *Execution time was taken into account*					
Min	12	0	10	7	5
Ave1	12	3	12	2	8
Ave2	12	6	9	4	6
Max	12	10	9	4	6
Average (Table 5.6)	12	4.75	10	4.25	6.25
(b) *Execution time was not taken into account*					
Min	0	0	2	0	2
Ave1	0	0	3	0	4
Ave2	0	0	0	4	3
Max	0	4	0	4	3
Average (Table 5.6)	0	1	1.25	2	3

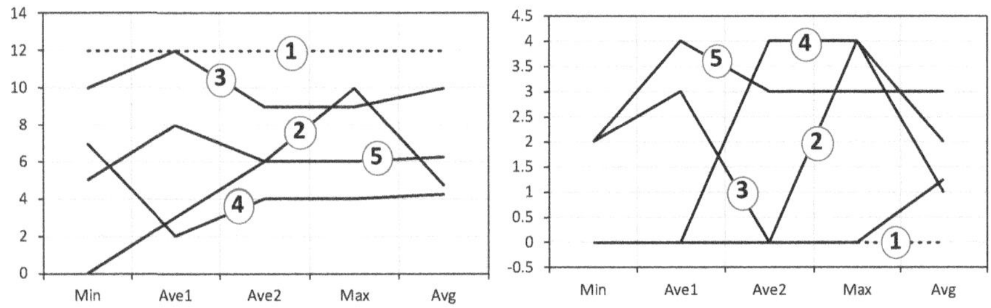

(a) Accounting for execution time (b) Without accounting for execution time

Legend: 1 – baseline; 2 - Sørensen-Dice; 3 – Jaccard; 4 – Jaro; 5 – Jaro-Winkler; rounded dashed rectangle – averages for all the thresholds

Fig. 5.25 Performance indices of the evaluated SSMs per terms similarity threshold with (**a**) and without (**b**) taking into account their execution time ranks

5.7 Validity and Scalability of the Optimized Term Extraction Pipeline

In this section, the evaluation of our optimized terminological saturation measurement and detection pipeline[27] is reported. The pipeline comprises the method (Sect. 3.6) and several relevant algorithms (Sect. 4.6) implemented in the corresponding software modules (Table 4.6). Section 5.7.1 is focused on the specifics of the experimental settings. Section 5.7.2 presents and discusses the results of the experiments.

5.7.1 Particularities in Experimental Settings

The objectives of our experiments with the optimized processing pipeline were to check its efficiency, domain neutrality, and effectiveness. With respect to the check of **efficiency**, we aimed, in the first experiment, to compare the baseline and optimized processing pipelines by verifying:

- **Correctness**. Were the merged partial *C*-Values, computed using the optimized pipeline, similar to the *C*-Values computed by the baseline?
- **Execution time**. What was the difference in the computation times, between the conventional and optimized pipelines, spent for the extraction of the bags of retained significant terms, in the same iterations?

[27] The material of this section is based on our publication [3].

Verifying correctness also aimed at validating the hypothesis $h1$, put forward in Sect. 3.6. The validity of this hypothesis, if proven experimentally, would have accomplished the proof of Theorem 3.5. As a straightforward practical consequence, the proof of the validity of $h1$ would have opened the way to use the optimized processing pipeline for extracting terms and measuring terminological saturation in document collections. The **efficiency** was assessed by the comparison of the execution times between the baseline and optimized pipelines on the same datasets.

In the second experiment. the **domain neutrality** of the optimized pipeline was evaluated. For that, the efficiency experiment, without measuring run times, was repeated for the other collection of documents that represented a different subject domain.

The third experiment aimed at evaluating the capability of the optimized pipeline to **effectively** waive the constraints of the baseline on the volume of datasets and, hence to be capable to process the document collection of industrial-scale volume.

In the first experiment, (**efficiency**) the DMKD collection (Sect. 5.2.2) was used. In addition to the available incremental datasets for the baseline processing pipeline (Sect. 5.2.2), the partition datasets were generated. In order to make sure that the data was compatible for both cross-evaluated processing pipelines, the same dataset increment ($inc = 20$) was chosen for the partition datasets. Hence, based on the available textual documents, 15 datasets of inc size forming the partition $\left\{D_i = \{d_j\}_{j=1}^{20}\right\}_{i=1}^{15}$ [28] were generated using our Dataset Generator software (Sect. 4.8).

The TIME collection (Sect. 5.2.2) was used for providing data for the **domain neutrality** experiment. Using TIME, 22 datasets of $inc = 20$ size forming the partition $\left\{D_i = \{d_j\}_{j=1}^{20}\right\}_{i=1}^{22}$ [29] were generated.

Finally, the KM collection (Sect. 5.2.2) was used for the experimental check of the **effectiveness** of the optimized pipeline. In this case, we have chosen $inc = 100$[30] for generating 75 datasets that formed the partition $\left\{D_i = \{d_j\}_{j=1}^{100}\right\}_{i=1}^{75}$.

The **descending citation frequency** (**dcf**) ordering (Sect. 5.5) of documents for generating partition datasets was used in all the experiments.

The execution flow, in our experiments on correctness, efficiency, and domain neutrality assessments, was configured in two parallel processing branches, as shown in Fig. 5.26—baseline (at the left) and optimized (at the right).

The baseline processed incrementally extended datasets as the inputs in iterations. The bags of term candidates were extracted by the UPM Term Extractor software (Sect. 4.8) into B_i. The run times ($time_i$) were measured for the baseline pipeline in every iteration.

---

[28] The partition of the DMKD-300 collection: https://github.com/OntoElect/Data/blob/master/DMKD-300-DCF-Part.zip.

[29] The partition of the TIME collection: https://github.com/OntoElect/Data/blob/master/TIME-DCF-Part.zip.

[30] The increment of 100 papers was chosen as appropriately granular for the shape of the diagrams in the figures.

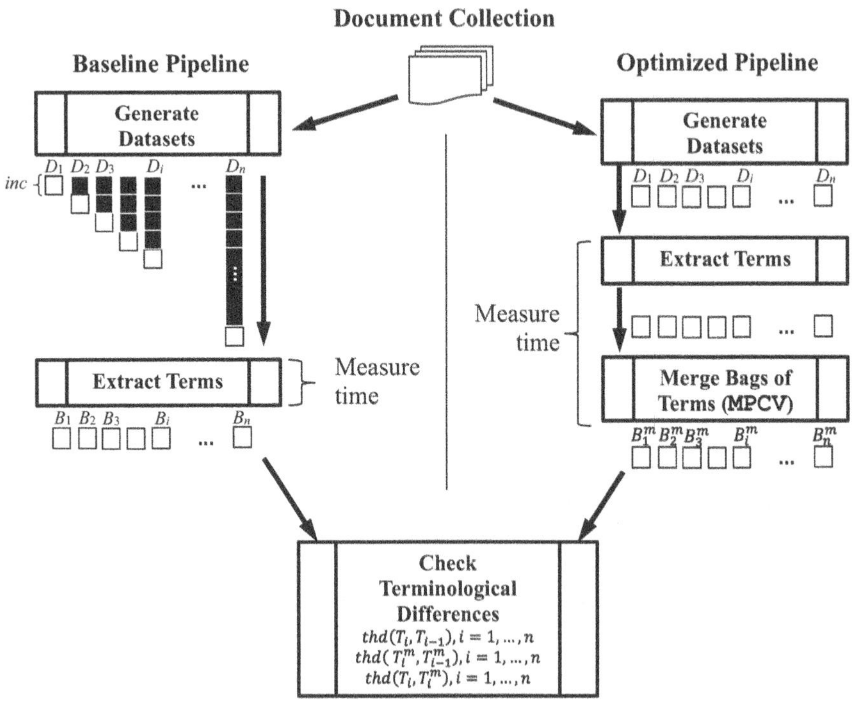

Fig. 5.26 The execution flow for the cross-evaluation of the baseline and optimized processing pipelines

The partition datasets (D_i) were processed using the optimized pipeline. The datasets D_i were fed into the **AC-CV** module, that retained significant terms with their partial C-Values into B_i The run times for terms extraction ($time_i^e$) were measured. At the next step, the (partial) bags of term candidates were iteratively merged by the **MPCV** algorithm (Sect. 4.4.2), producing the merged bags of term candidates ($B_i^m = \bigcup_{k=1}^{i} B_i$). **MPCV** module also measured the run times ($time_i^m$) for the merge operations. The total run times ($time_i = time_i^e + time_i^m$) were finally computed.

At the final step (bottom of Fig. 5.26), terminological difference measurements were done for

(a) The baseline—$thd(T_i, T_{i-1}),\ i = 1, \ldots, 15$ were measured,
(b) The optimized pipeline—$thd\left(T_i^m, T_{i-1}^m\right),\ i = 1, \ldots, 15$ were measured,
(c) The comparison of the baseline and optimized pipeline outputs—$thd\left(T_i, T_i^m\right),\ i = 1, \ldots, 15$ were measured.

Terminological differences $thd\left(T_i, T_i^m\right),\ i = 1, \ldots, 15$ were used to check the validity of $h1$. The rationale behind this check was that, if $h1$ held true, the terminological differences between the bags of retained significant terms T_i, coming from the baseline, and T_i^m, coming from the optimized pipeline, should have been negligibly small if not equal to zero. This check has been done using the baseline **THD** algorithm.

For checking **effectiveness**, the optimized pipeline (right part in Fig. 5.26) has been applied to the KM collection.

5.7.2 Results and Discussion

The results of the experiment that evaluated **correctness** and **execution time** are presented in Table 5.8 and visualized in Figs. 5.27 and 5.28.

It has been testified in the experiment on correctness that:

(a) The measurements of terminological differences for the results generated by the baseline (column 3 in Table 5.8 and curve (1) in Fig. 5.27a) and optimized pipeline (column 4 in Table 5.8 and curve (2) in Fig. 5.27a) have very similar values—the saturation curves in Fig. 5.27a overlap up to the very last two iterations. This difference at the tails of the curves could be explained by the fact that the pipelines accumulate regular noise slightly differently. It could be noted, as a side effect, that the optimized pipeline accumulated less regular noise than the baseline.

Table 5.8 *thd* measurements and run times for the baseline and optimized pipelines on DMKD

Iteration No. (i)	*Eeps*	*Tthd*			Run times (s)	
		T_i, T_{i-1}	T_i^m, T_{i-1}^m	T_i, T_i^m	B_i	B_i^m
1	2	3	4	5	6	7
1	12.00	62.89	62.89	0.00	30.41	31.94
2	15.50	31.51	32.22	3.64	60.47	32.79
3	18.00	23.38	22.17	2.85	85.34	31.64
4	19.65	18.04	17.63	3.08	111.72	31.85
5	23.22	15.80	15.57	3.31	147.30	37.11
6	24.00	10.07	9.50	3.86	153.78	32.25
7	24.00	10.67	10.14	3.40	195.27	41.49
8	26.00	9.14	9.19	5.28	217.62	43.28
9	28.00	8.68	8.54	6.95	268.59	47.16
10	28.53	7.56	7.42	5.21	296.49	41.32
11	28.53	7.30	6.44	6.13	324.39	41.58
12	28.53	6.53	6.56	5.88	363.05	45.20
13	30.00	6.43	5.42	10.07	401.94	40.55
14	38.00	15.47	3.13	9.09	401.19	39.81
15	38.00	3.53	15.37	6.14	459.75	50.12

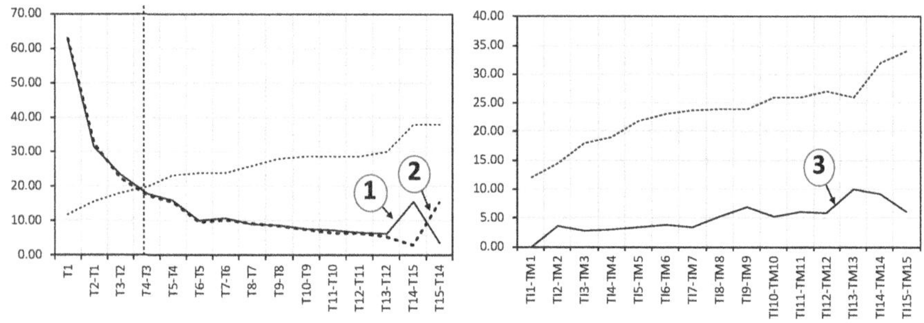

(a) Saturation measurements (b) Terminological differences

Legend: 1 – $thd(T_i, T_{i-1})$ for T_i computed using the baseline pipeline (C-Values); 2 – $thd(T_i^m, T_{i-1}^m)$ for T_i^m computed using the optimized pipeline (merged partial C-Values); 3 – $thd(T_i, T_i^m)$ for the bags of retained significant terms computed by different pipelines; dashed curve – ***eps***; vertical dashed line – saturation point

Fig. 5.27 The comparison of the outputs of the baseline and optimized pipelines for DMKD

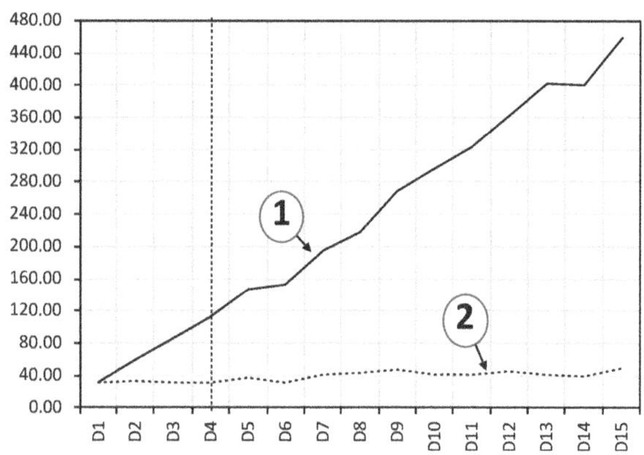

Legend: 1 – times (*sec*) taken for computing T_i using the conventional pipeline (C-Values); 2 – times (*sec*) taken for computing T_i^m using the optimized pipeline (merged partial C-Values); vertical dashed line – saturation point

Fig. 5.28 Run times of the baseline and optimized pipelines

(b) Terminological differences $thd(T_i, T_i^m)$ between the bags of retained significant terms coming from the baseline and optimized pipelines respectively (column 5 in Table 5.8 and curve (3) in Fig. 5.27b) did not exceed approximately one third of the individual term significance threshold values (*eps*) in all measurement iterations.

The observations (a) and (b) experimentally prove[31] the $h1$ hypothesis.

[31] One might remark that the reported evaluation was just an experiment with one document collection. Hence, for a different document collection the results might have been different with respect to

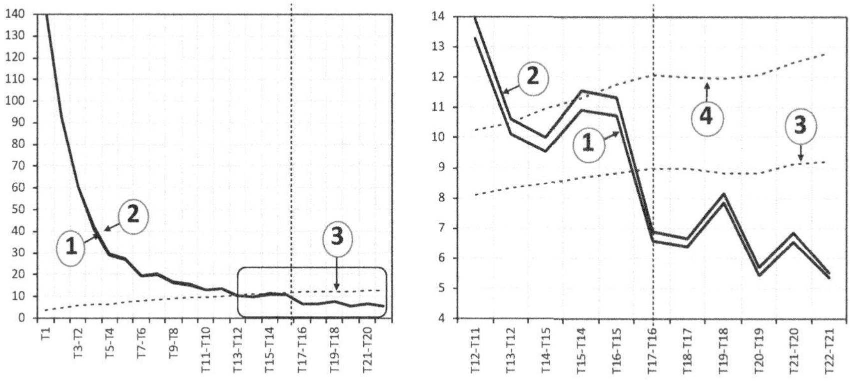

(a) Saturation measurements

(b) Saturation area in more detail – rounded rectangle in (a)

Legend: 1 – $thd(T_i, T_{i-1})$ for T_i computed using the baseline pipeline (C-Values); 2 – $thd(T_i^m, T_{i-1}^m)$ for T_i^m computed using the optimized pipeline (merged partial C-Values); 3 – eps_i^b computed for the bags of retained significant terms (T_i); 4 – eps_i^m computed for the bags of retained significant terms (T_i^m); vertical dashed line – saturation point

Fig. 5.29 Terminological difference between the bags of terms computed using the baseline and optimized pipelines on TIME datasets

The run times measured for the baseline and optimized pipelines are presented in columns 6 and 7 of Table 5.8 and as curves (1) and (2) in Fig. 5.28. Their comparison demonstrates that the optimized pipeline, which spent near to constant times in every iteration, outperformed the baseline significantly.

The results of our experimental check of the **domain neutrality** of the optimized pipeline are visualized in Fig. 5.29. Identically to the case of DMKD, the thd values measured for the outputs of the baseline and optimized pipelines, applied to the TIME datasets, were similar. However, the *eps* values computed using the bags of term candidates, coming from the baseline (eps^b) and optimized (eps^m) pipelines, were noticeably different. As it could be seen in Fig. 5.29b, the values of eps^b were 25–30% higher than the values of eps^m.

Furthermore and in difference to the DMKD experiment, $thd\left(T_i^m, T_{i-1}^m\right)$ for TIME appeared to be noticeably higher than eps_i^m. At the first superficial view, it looked like the baseline and optimized pipelines yielded different bags of retained significant terms in the case of TIME. It was hypothesised that the differences were caused by differently accumulated regular noise.

To check this hypothesis, the bags of term candidates B_{22} and B_{22}^m, extracted in the last iteration by the baseline and optimized pipelines respectively, were manually

---

the validity of $h1$. Our counter-argument is that the computation of merged partial C-Values, similar to the computation of C-Values, is collection- and domain-independent. We proved that in our domain neutrality experiment. Results are the same for the other document collection representing a different subject domain.

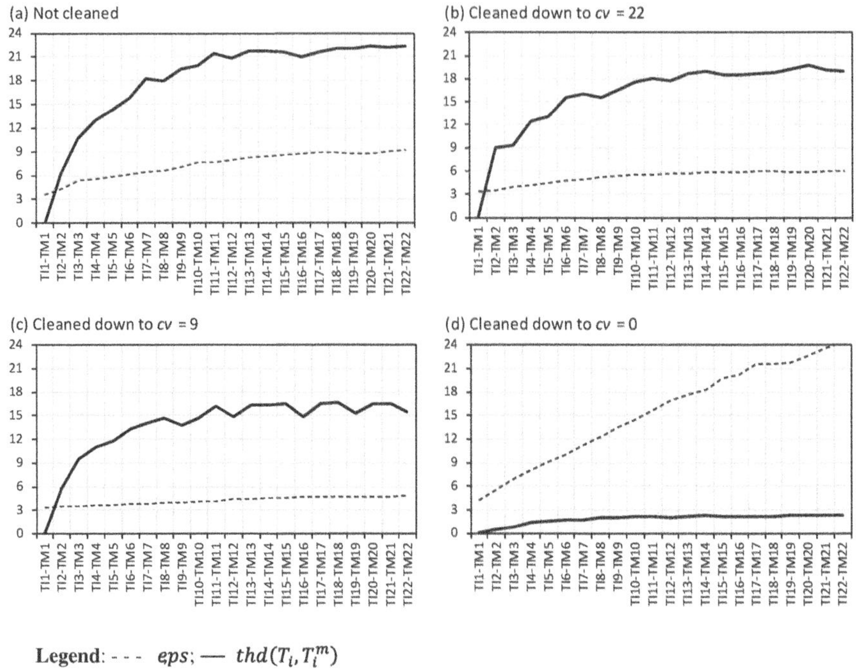

Fig. 5.30 The dynamics of regular noise influence on the terminological difference between T_i and T_i^m on TIME

checked. This inspection was performed by an expert in the subject domain of TIME and revealed that the false positives[32] resided at the tops of both B_{22} and B_{22}^m. The C-Values (or merged partial C-Values) of these false positives were often and on average higher than that of the true positives, as pictured in Fig. 5.31 for B_{22}^m, which was nothing else then the substantial accumulation of the regular noise.

To lower the impact of the accumulated regular noise, the false positives were manually extracted from B_{22}^m and then removed from all the bags of terms B_i and $B_i^m, i = 1, \ldots, 22$, using the stop-term remover module (Sect. 4.8). This procedure was applied in several iterations to observe the dynamics of the changes in $thd(T_i^m, T_{i-1}^m)$. In the first iteration, the false positives were removed down to their cv or $mpcv$ equal to 22. In the second iteration—down to cv or $mpcv$ equal to 9. In the last iteration—down to cv or $mpcv$ equal to 0. The results of measuring $thd(T_i^m, T_{i-1}^m)$ after noise removal in these iterations are pictured in Fig. 5.30b–d.

Figure 5.30d shows that, after the removal of all the accumulated regular noise from the sets of term candidates, $thd(T_i, T_i^m)$ became substantially lower than the corresponding values of eps. The situation with the de-noised TIME bags of term

[32] A false positive here and in Fig. 5.31 is a term candidate string with high C-Value, i.e. significance score, but appearing to be not a term for a human expert (domain knowledge stakeholder)—i.e. an accumulated regular noise instance.

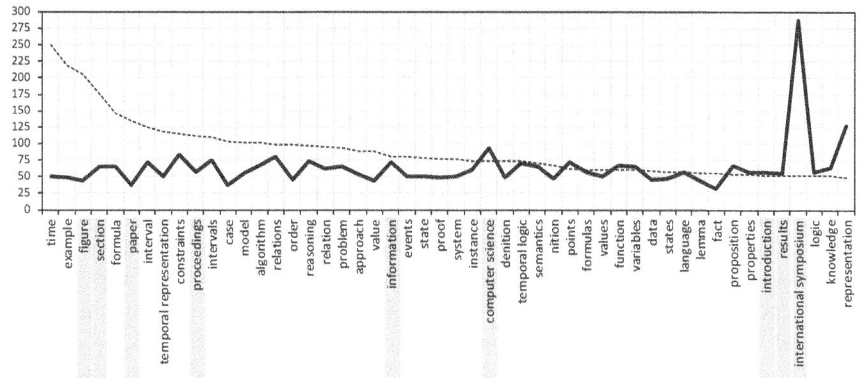

Legend: false positives that represent accumulated regular noise are greyed; dashed curve – $mpcv/20$; thick solid curve – $cv - mpcv$.

Fig. 5.31 Accumulated regular noise observed in the largest bag of terms (B_{22}^{m}) extracted by the optimized pipeline from TIME

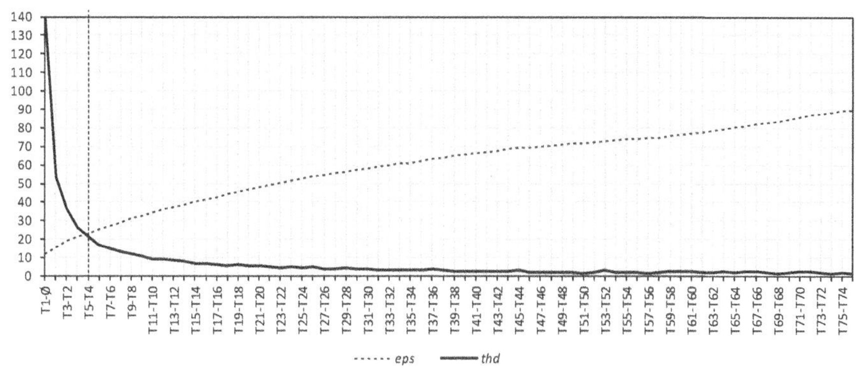

(a) Saturation measurements. Dashed curve – eps_i^m; solid bold curve – $thd(T_i^m, T_{i-1}^m)$; vertical dashed line – saturation point.

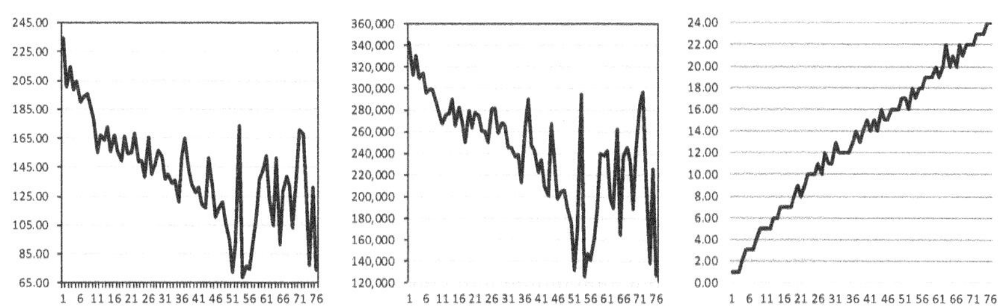

(b) Extraction time per dataset, *sec*

(c) No extracted term candidates per dataset

(d) Merge time per bag of terms, *sec*

Fig. 5.32 Saturation and runtime measurements for the use of the optimized pipeline to process the datasets of KM partition candidates (c.f. Fig. 5.30d) is better than that with DMKD bags of term candidates (c.f. Fig. 5.27b). The reason for this improvement was that DMKD bags of term candidates were not de-noised.

The results of the experiment on verifying the **effectiveness** of the optimized processing pipeline are presented in Fig. 5.32. In this experiment, the optimized pipeline was applied to process the partition datasets generated using the documents of the KM collection.

Our effectiveness experiment testified that the use of the optimized pipeline waives the volume constraint, caused by the specificity of the original *C*-Value method [7], regarding input datasets. Furthermore, the constraint of the *C*-Value method is removed with the help of our optimized approach in a broader context—covering all ATE tasks that use *C*-Values for scoring strings in the bags of extracted term candidates. Furthermore, the comparison of the measured run times (Fig. 5.32b) shows that the parallel processing of the partition datasets could facilitate making our approach to terms extraction from texts substantially more efficient and cost-effective than the original *C*-Value method. It is straightforward that parallel processing is possible due to the disjointness of the partition datasets. The only step that could not be run in parallel is merging partial *C*-Values. However, the runtime overhead for this step is negligible (Fig. 5.32d) compared to the times to be spent for extracting the bags of term candidates.

5.8 Summary

In this chapter, we focused on meeting our research objective **RO3** (c.f. Sect. 2.13). For that, we set up our series of experiments, performed the experiments, discussed the results, and drew several recommendations.

For the experimental set-up, we formulated our experimental objectives and planned our experiments in a way to meet these objectives. The objectives were formulated in Sect. 5.1 in a way to frame these out by the evaluation and validation facet of the research questions **RQ9**, **3**, **1**, **8**, and **7** (c.f. Table 2.5). As the parts of the set-up and plan, we also: (i) developed the experimental workflow; (ii) collected the required data; (iii) specified the measurable aspects and measures for assessing the results; and (iv) configured the environment for experimental runs. Experimental set-up was elaborated in Sect. 5.2.

In Sect. 5.3, we answered **RQ9** about the correctness of the developed method and algorithmic suite in the boundary cases. For that, we performed the experiments using the synthetic collections, 1DOC and RAW, that had been designed to represent the data for the boundary cases: quick and steady saturation and no saturation. The results of the experiments demonstrated that the method and its algorithmic and software implementations are correct as produce expected results.

In Sect. 5.4, we answered **RQ3** about the choice of the most appropriate ATE software tool for terminological saturation measurement. As a result, UPM Term Extractor software has been selected based on the cross-evaluation against

NaCTeM TerMine. Therefore, UPM Term Extractor has been further used in all the computations as a part of the baseline computation pipeline.

In Sect. 5.5, we answered the question about the impacts of different possible document orderings on terminological saturation measurement (**RQ1**). Based on cross-evaluating five different orderings against the measurable aspects (Table 5.2), we recommended the descending citation frequency (**dcf**) as the most balanced across different aspects and subject domains.

In Sect. 5.6, we evaluated if the use of similar terms grouping has positive impact on terminological saturation (**RQ8**). In this evaluation, we applied the refined computational pipeline, with term grouping and terms similarity measurement to three different real document collections across subject domains. In the experiments, we vary the methods for term similarity measurements and corresponding thresholds in order to find the most productive configuration in terms of the quality of the result and computational efficiency. As a result, we recommended two best-ranked configurations for use.

Finally, in Sect. 5.7, we answered the question about the validity, efficiency, and scalability of the optimized computational pipeline (**RQ7**), developed in Sects. 3.6 and 4.4. We have proven, experimentally, the hypothesis $h1$ of Theorem 3.5, which accomplished the proof of this theorem. We also found out that the optimized pipeline is computationally efficient, straightforwardly parallelizable, and scales well enough to waive the inherent memory constraint of the State-of-the-Art ATE implementations. Hence, it could be used for processing industrial size document collections.

References

1. Ermolayev V, Batsakis S, Keberle N, Tatarintseva O, Antoniou G (2014) Ontologies of time: review and trends. Int J Comput Sci Appl 11(3):57–115
2. Kosa V, Chugunenko A, Yuschenko E, Badenes-Olmedo C, Ermolayev V, Birukou A (2017) Semantic saturation in retrospective text document collections. In: Proceedings of ICTERI 2017 PhD symposium. Kyiv, Ukraine, May 2017. CEUR-WS, vol 1851, pp 1–8
3. Kosa V, Chaves-Fraga D, Dobrovolskiy H, Ermolayev V (2020) Optimized term extraction method based on computing merged partial *C*-values. In: Ermolayev V et al (eds) Revised selected papers of ICTERI 2019. Communications in computer and information science, vol 1175. Springer, Cham, pp 24–49. https://doi.org/10.1007/978-3-030-39459-2_2
4. Kosa V, Chaves-Fraga D, Naumenko D, Yuschenko E, Badenes-Olmedo C, Ermolayev V, Birukou A (2018) Cross-evaluation of automated term extraction tools by measuring terminological saturation. In: Bassiliades N et al (eds) Revised selected papers of ICTERI 2017. Communications in computer and information science, vol 826. Springer, Cham, pp 135–163. https://doi.org/10.1007/978-3-319-76168-8_7
5. Kosa V, Chaves Fraga D, Naumenko D, Yuschenko E, Badenes-Olmedo C, Ermolayev V, Birukou A (2017) Cross-evaluation of automated term extraction tools. Technical report TS-RTDC-TR-2017-1, 30 Sept 2017, Dept of Computer Science, Zaporizhzhia National University, Ukraine. http://ermolayev.com/TS-RTDS-TR-2017-1.pdf. https://doi.org/10.13140/RG.2.2.31187.07207

6. Kosa V, Chaves-Fraga D, Naumenko D, Yuschenko E, Moiseenko S, Dobrovolskyi H, Vasileyko A, Badenes-Olmedo C, Ermolayev V, Corcho O, Birukou A (2018) The influence of the order of adding documents to datasets on terminological saturation. Technical report TS-RTDC-TR-2018-2-v2, 21 Nov 2018, Dept. of Computer Science, Zaporizhzhia National University, Ukraine. https://doi.org/10.13140/RG.2.2.28382.54086
7. Frantzi KT, Ananiadou S (1999) The *C*-Value/NC-Value domain independent method for multi-word term extraction. J Nat Lang Process 6(3):145–179. https://doi.org/10.5715/jnlp.6.3_145
8. Kosa V, Chaves-Fraga D, Keberle N, Birukou A (2019) Similar terms grouping yields faster terminological saturation. In: Ermolayev V et al (eds) Revised selected papers of ICTERI 2018. Communications in computer and information science, vol 1007. Springer, Cham, pp 43–70. https://doi.org/10.1007/978-3-030-13929-2_3

Chapter 6
Saturated Terminology Extraction and Analysis in Use

The method and software implementation of saturated terminology extraction has already been used in external industrial and academic projects. This Chapter reports about one industrial (Sect. 6.1) and one academic (Sect. 6.2) use case of our research contribution. These use cases are the early and ongoing attempts to check if the results of our research could be potentially useful in external industrial and academic settings. Hence, the currently reported outputs of the presented use cases have to be regarded as potential utility and benefits indicators. Consequently, Sect. 6.3 outlines our deliberations about the practical implications, including limitations and benefits, which might be expected by the potential adopters of our research output. Section 6.4 presents our prognosis about the potential scope of the technology transfer to the industry of scholarly publishing, as a focal sector. Section 6.5 summarizes use experience and the estimated prospects for technology transfer.

6.1 Checking Gartner Trend Prediction

This Section, reports the use of our method for terminological saturation detection and measurement for terminology analysis in the ongoing industrial project[1] performed in cooperation with Group BWT.[2]

Every information technology provider, be it a start-up company, small or medium-sized enterprise, or a bigger corporation, observes and monitors the evolution of technology and application trends in their foci of operation and interest. This activity is essential for the survival of their business, as it directly affects their prospects to improve their competitiveness in the focal markets.

[1] SAGOIT-IT: Strategic Analysis of R&D Gaps and Opportunities for Industrial Uptake in Trending IT Fields –the project funded by Group BWT, llc.

[2] https://groupbwt.com/.

V. Kosa and V. Ermolayev, *Terminology Saturation*, Cognitive Science and Technology,
https://doi.org/10.1007/978-981-16-8630-6_6

One of the worldwide reputed sources of information on technology trends in this industrial sector is the collection of the reports published annually by Gartner.[3] Their methodology is to interview the key professionals in the relevant fields and present the results of the analytical processing of these interviews in a series of reports. Each of these reports targets its own focal audience belonging to senior management. Relevant Gartner reports on technology trends are [1, 2]. Based on the information provided by these reports, technical and technology management in a company might assess their current position, available capabilities and competencies, and potential to follow the focal trend and thus increase their competitiveness in the mid-term perspective.

However, Gartner's qualitative research output on technology trends is too high-level and lacks detail for properly positioning, directing, and improving the research and development (R&D) agenda at a company level. Hence, having a vision of a trend from Gartner and having selected a priority technology, a company needs to pay a closer and more detailed technical look at the chosen context.

Such a closer look into the lower-level technology potential analysis is presented further in this Section, with regard to one chosen trend among those outlined by Gartner [1]—the Generative Adversarial Networks (GAN) technology.[4] Our approach in this use case was based on the terminological analysis of the collected document corpus for the subject domain of GAN. For finding out if there was an opportunity window within the chosen focus of interest, terminological differences between academic, industrial, and collaborative subsets of the retrieved collection of documents were measured. The intuition behind this partitioning was threefold:

- Terminological footprints extracted from a representative set of relevant documents reflect the State-of-the-Art and State-of-the-Technology in a field. Hence, if a stable and mature terminological footprint is observed for the focal field of study this field could be regarded as technologically mature.
- The documents authored by academics are obviously at the forefront of research; the adoption of technology by industry follows academic advances with a time lag. Hence, the differences between academic and industrial terminological footprints indicate the lag in technology maturity. Furthermore, the absence of terminological saturation in the term-set extracted from industrial papers indicates that the industry is not mature in adopting the technology, at least that far.
- The analysis of the terminological footprint of collaborative papers is a good way to look at how active industry is in the research related to the selected technology and also how eager it is to adopt the technology from academy.

[3] https://www.gartner.com/en.

[4] The detailed presentation of our results in this study is available as a technical report [20].

6.1.1 *Questions and Method of the Study*

To have a comprehensive and sufficiently detailed understanding of the potential of the technology under analysis to influence the markets and offer competitive advantage for an early technology adopter, the answers to the following important questions are of particular relevance:

- **Q1**. Are the research contributions and respective professional community around a technology mature?

The answer to this question is important because maturity: (i) implies the availability of a well-shaped body of knowledge about the State-of-the-Art in the field and established research community behind this field; and (ii) facilitates to devising a clear identification of the R&D gaps of potential interest.

- **Q2. Are there any gaps in the knowledge about the technology between academic researchers and industrial adopters?**

The answer to this question is important because, if the gaps are identified, the objectives to narrow or close these gaps could be more easily identified in the company's R&D agenda.

- **Q3. What is the temporal lag between the identified State-of-the-Art and the current State-of-the-Technology?**

The answer to this question is important for finding out if there is still a time and market window for a technology adopter to make use of the technology in their business.

To answer the questions **Q1** to **Q3**, the terminological footprints of the documents (scholarly papers) that describe the subject domain (GAN) chosen for our analysis are examined. These terminological footprints are extracted for different sub-groups of documents:

- Authored by academics
- Authored by industrialists
- Authored collaboratively by academics and industrialists.

The terminological footprints of these sub-groups of documents are compared using the terminological difference function (*thd*) (Sect. 3.3). Based on these comparisons, the answers to the questions are produced.

For ensuring that the comparisons of the footprints are grounded, all the collected documents have to be relevant to the focal subject domain and the collection has to be representative. Relevance in our methodology is assured by the use of the method of document sampling [3]. Representativeness is checked by measuring terminological saturation in the sample of documents, using the method and software presented in Chaps. 3 and 4.

The method for documents sampling [3] applies the combination of snowball sampling, probabilistic topic modelling, and citation network analysis to search

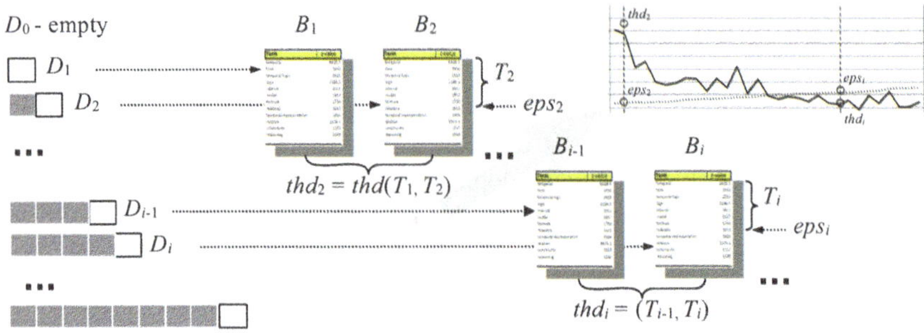

Legend: D_i – plain text datasets; B_i – bags of extracted terms;
eps_i – individual term significance threshold for B_i;
T_i – bags of retained significant terms;
thd_i – terminological difference between T_{i-1} and T_i;
i – the number of iteration

Fig. 6.1 Terminological difference measurement flow for detecting saturation in a part of the collection

through the titles, keywords, and abstracts of scientific papers indexed by Microsoft Academic.[5] It requires that a few relevant and significant documents (a seed) be provided to kick-start the iterative process. As seed documents, the method recommends providing frequently cited review papers published later than five years from the present.

The retrieved document collection is split into three parts, based on the information on the affiliations of the authors of individual papers. In this project, the split has so far been done manually by the collection curator.

The configuration of the pairs of the bags of terms used in terminological difference measurements is different for answering different questions, as explained below.

To answer **Q1**, terminological saturation analysis is performed individually on each of the three collection parts. The computation flow for these measurements is outlined in Fig. 6.1.

For answering Q1, the sequence of the pairs of the retained significant terms ($T_{i\text{-}1}$, T_i), $i = 1, \ldots, N/inc$ is used, where N is the number of documents in a document collection part, $inc = 10$.

To answer **Q2**, terminological differences are measured between the bags of retained significant terms, extracted from different parts of the collection: Academic and Industrial; Academic and Collaborative; Industrial and Collaborative—at the same iteration of the successive approximation process of measuring terminological saturation. The computation flow for these measurements is outlined in Fig. 6.2 for the pair of academic and industrial parts.

---

[5] https://academic.microsoft.com/.

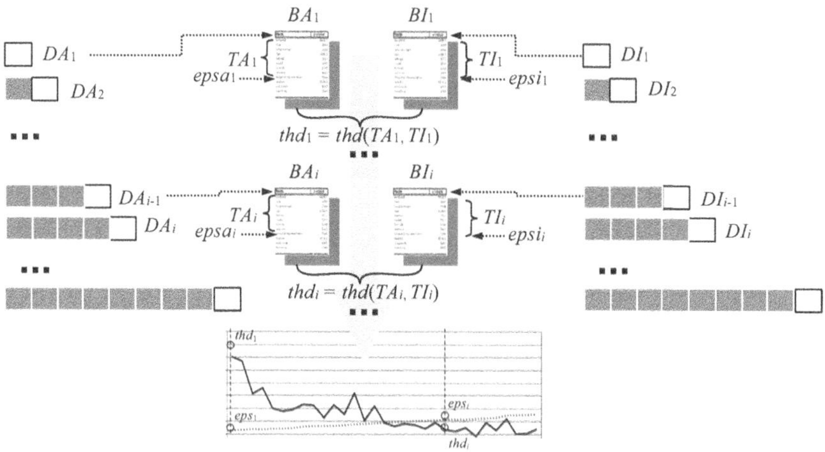

Legend: DA_i, DI_i – plain text datasets of different collection parts (*A* – academic, *I* - industrial);

BA_i, BI_i – bags of extracted terms from different datasets;

$epsa_i$, $epsi_i$ – individual term significance thresholds for BA_i, BI_i;

eps_i – $\min(epsa_i, epsi_i)$

TA_i, TI_i – bags of retained significant terms from different datasets;

thd_i – terminological difference between TA_i and TI_i;

i – the number of iteration

Fig. 6.2 Computation flow for assessing terminological differences between the bags of retained significant terms of different collection parts

Finally, to answer **Q3** in the future work, it is planned to perform retrospective terminological difference analysis. The retrospective is based on the assumption that there is a lag of several years between an academic contribution (presented in academic publications) and effective industrial adoption of this contribution (presented in industrial publications). Straightforwardly, the adoption happens later. Hence, industrial papers might operate the set of domain-specific terms that is very similar to those of the academic pool of papers, but several years before. This assumption is indirectly supported by Gartner, who predicts, for each reviewed technology, how many years it will take to reach the plateau of productivity. For example, GAN is projected [1] to reach the plateau in 5–10 years from 2019.

One of our objectives in this ongoing project is to find out what is the length of this time lag, in years, which is the indicator of the breadth of the opportunity window with respect to the adoption of the technology. For doing that, several snapshots of the academic partition of the retrieved collection are prepared. These snapshots contain papers published: up to date; up to one year ago; up to two years ago, ..., up to ten years ago. Terminological differences are then examined between each of the prepared academic part snapshots and the industrial and collaborative parts, as it has been done while answering **Q2**. If there is an academic snapshot, which is terminologically the closest to the industrial or collaborative part, and the terminological distance is lower than *eps*, then the time lag is determined for the pair.

6.1.2 Experimental Results and Discussion

As coined out by Goodfellow et al. [4], GAN is a deep learning approach in which two models are trained in parallel and in a competitive way. GANs are currently one of the mainstream approaches in Deep Learning, with a handful of different computer vision, speech processing and recognition, natural language processing applications.

The sampling of the papers relevant for GAN was kicked-off by feeding the two recent reviews by Gui et al. [5] and Pan et al. [6] to the processing pipeline as the seed. Based on this seed, 262 documents were retrieved and split into parts: (i) Academic—138 papers; (ii) Industrial—38 papers; and (iii) Collaborative—86 papers. Using the datasets generated from these collection parts, terminological saturation measurements and analysis were performed. It has been found out that the sample was not representative. To improve on representativeness, the seed set of papers was extended by the first ten most frequently cited documents from the retrieved sample. These papers were: [7]:17,224.75 (ind)[6]; [8]: 6,216.71 (acad); [9]: 3,971.83 (clb); [10]: 3,427.66 (ind); [11]: 3,011.00 (acad); [12]: 2,965.00 (acad); [13]: 2,353.00 (acad); [14]: 2,020.57 (acad); [15]: 1,913.40 (acad); and [16]: 1,432.71 (acad).

The second sampling round, which used the extended seed, resulted in collecting 617 documents in total. The collection was split as: (i) Academic—319 papers; (ii) Industrial—100 papers; and (iii) Collaborative—198 papers. These parts were used in the second iteration of terminological analysis, which allowed us to answer the questions **Q1** and **Q2** of the study.

To answer the **Q1** question about the maturity of the body of published knowledge on GAN, terminological saturation for all three collection parts was measured. The results of these measurements are visualized in Figs. 6.3 and 6.4.

Only one part of the three—the Academic—demonstrated terminological saturation (Fig. 6.3a). This fact indicated the maturity of the terminological set in this part and therefore, of the academic community in GAN. The other two parts—Industrial and Collaborative—did not demonstrate terminological saturation (Fig. 6.3b, c) based on the collection of papers that had been retrieved.

The low volatility of terminological differences measured for the academic part (Fig. 6.4a) confirmed that detected saturation is stable—i.e. it would have most likely persisted if additional academic papers on GAN were added to the collection. Indeed, the range (minimal to maximal) of volatility values in the estimated saturation zone (for the pair of the bags of retained significant terms (17, 18) and onwards) is 5.2832 + 5.8326 = 11.1158, which is below the individual term significance threshold:

[6] Here and below in the list of the seed papers, the number after the reference is the number of citations per year (Google Scholar) followed by the collection part identifier (ind/acad).

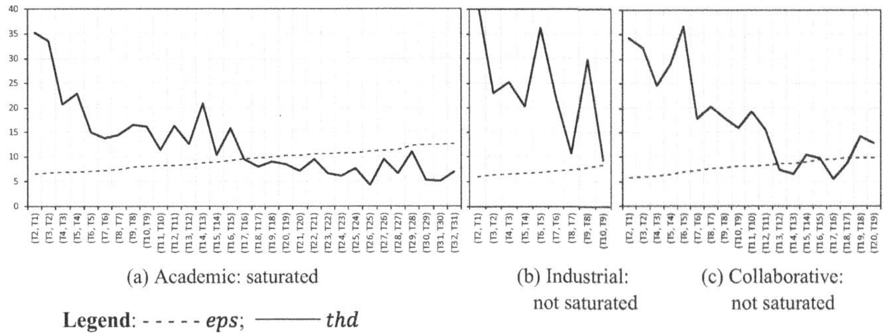

Fig. 6.3 Terminological saturation measurements for the Academic, Industrial, and Collaborative parts of the GAN collection

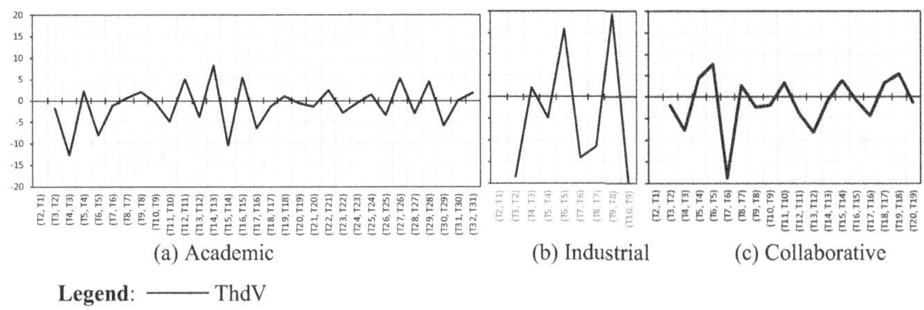

Fig. 6.4 Terminological difference volatility measurements for the Academic, Industrial, and Collaborative parts of the GAN collection

11.3329 at (27, 26) growing to 12.7122 at (32, 31) with the trend of monotonic growth.

To answer the **Q2** question about the gaps in the body of published knowledge on GAN, terminological differences were measured in the three pairs of collection partitions: (i) Academic and Industrial; (ii) Academic and Collaborative; and (iii) Collaborative and Industrial. The results of these measurements are visualized in Fig. 6.5.

The curves in Fig. 6.5 visualize the measured terminological differences, in the published papers on GAN, between the academic and industrial parts of the collection. Even in the case of counting in the collaborative work, these terminological differences were substantial. Hence, the gaps between the bodies of published knowledge in different parts of the collection were also substantial. These point out that the State-of-the-Art and State-of-the-Technology were different. Therefore, there is a broad opportunity window for the technology adopters to benefit from the effort on GAN transfer to industry.

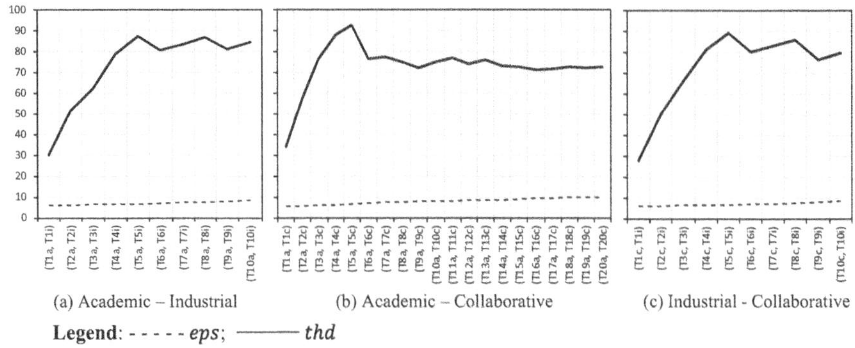

Fig. 6.5 Terminological difference curves for the pairs of collection parts

6.2 Instrumenting the Literature Review Activity of Master Students

This Section reports the use of our method for terminology saturation detection and measurement in the academic curriculum of the International Master Programme on Computer Science and Data Science of Ukrainian Catholic University.[7] It has been used as a part of the instrumental pipeline offered to the students to help them collect their literature samples for the reviews of the related work in their Master theses. Didactically, this work has been offered as the first course assignment within the course on academic writing in the fall semester of their first year of studies.

6.2.1 Task for Students

At the lectures, the students were informed that a good review of the related work is an essential ingredient for any research project to arrive at understanding the gap in the published results by the others and, hence, the motivation to push forward the State-of-the-Art in the chosen research field. This course assignment was aimed at giving the students an opportunity (and guidance) to elaborate this component for their Master project in a systematic, hence, effective and efficient way.

The assignment was offered after the completion of the first part of the course. This assignment dealt with conducting literature search and selection within the scope of the chosen topic of a Master project. After the literature sample had been collected, the chosen literature sources were read, reviewed, and analysed with the objective to identify a research gap and motivate themselves to narrow this gap. The findings had to be documented in a Technical Report.

[7] https://ucu.edu.ua/en/.

The report was evaluated based on the aspects that resemble the criteria used for the evaluation of review papers. These aspects and corresponding grading scales were:

- Methodology (0–5)
- Representativeness (0–10)
- Relevance (0–10)
- Structure (0–5)
- Gap analysis (0–10)
- Writing style (0–5).

The students were advised that, as a pre-requisite for making their review and analysis grounded and unbiased, the collection of the literature sources had to be:

- **Representative**, which is sufficiently complete, to date, for covering all the significant research and development accomplishments in the chosen field
- **Relevant**, which is containing only the sources, that, in their substance but not by the title, report the results that might be of use for or influence the work planned in their Master project.

With respect to the **representativeness** of the collected literature, the students were suggested to prove objectively that, at least, all the mainstream related work to date was covered by the collected literature sample.

With respect to answering the **relevance** question in a cogent way, the student had to have a well-formed idea of what is planned to be done in the project, which could be refined in iterations based on the discovered literature collection of the previous iteration (the Active Learning approach [17]). The students have also been provided with instrumental support for this iterative activity of literature sampling.

The provided instrumental pipeline was targeted to help them answer the representativeness and relevance questions as follows:

- **Relevance**. It was suggested to provide a small "seed" set of research papers that are indexed by Microsoft Research.[8] The metadata of these papers was further fed into the Controlled Snowball Sampling tool [3]. The tool generated the catalogue of the selected relevant publications using citation network analysis, probabilistic topic modelling, and snowball sampling. The full texts of the papers that were publicly available were downloaded for further reading and analysis. Iterations could have been done via refining the "seed" set of research papers based on the result of the previous iteration.
- **Representativeness**. It was suggested to use the collection catalogue and downloaded full texts of the publications for conducting the analysis of terminological saturation [18] within the collected set of papers. Terminology saturation in the subset of papers, if observed, could have been used as proof of the representativeness of this subset. Furthermore, the saturated subset was smaller than the entire collected set of papers and required less effort for reading and analysis.

---

[8] https://academic.microsoft.com/.

Both tools were bundled in the computation pipeline as presented in Sect. 4.1 (Fig. 4.3).

Students were not requested to use the offered instrumental bundle. Instead, they were free to choose any other means to collect the representative and relevant set of publications for their literature analysis. They were informed, however, that the objective arguments for proving representativeness and relevance still had to be provided in the technical report to earn appropriate aspect grades. They were also requested to present their chosen methodology for literature collection in a manner and detail that allowed the reproduction of this work.

6.2.2 *Method Adoption Results*

Nineteen out of 23 students in the class have submitted their course assignment reports. Among those 19:

- Sixteen students have used the suggested instrumental pipeline in part (for literature sampling), though it was not obligatory
- Four students used the suggested instrumental pipeline in full (for literature sampling and terminology saturation analysis). Those four all appeared in the highest-grade quartile after all the reports had been evaluated by the lecturer.

Based on these trials, it could be stated that the technology and software pipeline for terminology saturation measurement and analysis was easily adoptable by master-level students and was perceived by them as useful.

6.3 Practical Benefits and Limitations

From the practical viewpoint, there is a noticeable potential benefit in using a compact and stable saturated term-set and corresponding terminological core sub-collection. Firstly, text and data mining of the full-text papers for commercial purposes incurs additional costs.[9] Therefore, buying access to a terminological core sub-collection is, clearly, times more efficient than to the complete collection. Secondly, the successive approximation approach (c.f. Chap. 3) allows purchasing the increments, in chunks, until the terminological core sub-collection is assembled. This allows paying exactly for the minimal effective set of papers.

Furthermore, compared to the State-of-the-Art results (c.f. Chap. 2), the use of our method for extracting terminological core sub-collections and respective saturated term-sets brings the following additional advantages:

[9] For example, the policies for the use of Springer-Nature text and data mining API could be accessed at https://www.springernature.com/gp/researchers/text-and-data-mining.

- **Domain neutrality**. In difference to several machine learning-based solutions for ATE, our method neither needs labelled data for training nor requires to be retrained when switching between subject domains. Hence, it is fully domain neutral, yet delivers the ATE results of good quality. Therefore, a potential adopter of our technology is able to apply it seamlessly on any text collections in any domain of their interest without adaptation expense.
- **ATE method neutrality**. The method (Chap. 3) and software pipeline (Chap. 4) use, currently, the *C*-Value method for ATE. This does not constrain the technology, however, by binding it to the *C*-Value method, which could be easily replaced by a preferred alternative. The alternative has however to be implemented in a way to output the results in the same form as described in Sect. 3.1. This means that every term in the extracted bags of terms ought to be represented by the pair $\langle t_i, \text{score}_i \rangle$. , where t_i is a character string representing the candidate term and score_i is e significance value of this term candidate. The possibility to alter the ATE method also opens the way to process texts in languages other than English. Furthermore, it enables an easy upgrade of the solution, by incorporating any novel high-quality ATE method that might become available in the future.
- **Full-text processing**. The State-of-the-Art approaches for ontology learning from texts that scale to industrial size collections process only the abstracts of the documents (c.f. Sect. 2.3). This is because: (i) the minimal representative subset of documents is not considered; and (ii) used ATE methods do not scale well with the growth of the volume of processed texts. Our method overcomes these two deficiencies of the State-of-the-Art, which allows it to cope with full-text processing in a better scalable way. Furthermore, the quality of term extraction from full texts is substantially higher than that from abstracts. Hence, our method provides better quality outputs for further use, for example in literature reviews or ontology learning from texts.
- **Waived restriction on collection size**. As elaborated in Sects. 3.6, 4.4, and 5.7, our optimized processing pipeline is not constrained by the volume of the text to be processed. Therefore, and in difference to the predecessor methods, it is capable to process industrial size collections of documents.
- **Easy parallelization**. In difference to the other ATE(-based) pipelines that require all the text (collection) to be present at one processing node, our optimized approach and implementation (Sect. 3.6 and 4.4) require only one partition dataset of the collection (inc documents) at a time. Hence, as proven in Sect. 3.6, one might use as many parallel nodes as available for concurrent processing of the partition datasets of a collection. As shown in our effectiveness experiment (Sect. 5.7), using this approach substantially reduces run times.
- **Balanced account for publication type and impact**. The use of the recommended **dcf** ordering (Sect. 5.5) ensures that the extracted terminologically saturated sub-collection contains all the texts having the accumulated impact that forms the majority opinion in the specialist community in the subject domain.

The solution for terminological saturation detection and measurement in its current shape, as presented in the book, is a research prototype. Therefore, it is not free from some limitations. The limitations have been mentioned throughout the book and could be overcome, at least in part, in our planned future work (c.f. Sect. 7.2). The limitations that have been currently identified are summarized as follows:

- **Only texts in English are processed**. Currently, and due to the use of the NLTK API[10] in the implementation of our ATE module, only texts in English could be processed by our pipeline. This limitation is not very strong as English is the first preferred language for the majority of professional documents, scholarly publications in particular. However, one may envision plenty of applications that require the extraction of saturated term-sets in languages other than English in industrial or academic practice. Furthermore, the processing of cross-language document collections could be practically significant. This limitation could be attacked by providing optional mono- or cross-lingual ATE modules as a part of the extension of the reported solution.
- **Terminological saturation could not be predicted**. After proving Theorem 3.3 (Sect.3.5), we acknowledged that knowing about the sufficient conditions for the existence of terminological saturation does not yet allow predicting that saturation will occur in a particular document collection. One has to go through the entire successive approximation process to find out if the conditions are met and terminological saturation detected. The method might become more efficient and effective if it allows to make the prediction of the existence of terminological saturation in the given document collection based on several initial iterations of successive approximations. This limitation of the method could be attacked by analysing the distributions of the terms in the few initial iterations, as proposed by Kosa and Ermolayev [19].
- **Use cases are insufficiently fully elaborated**. It becomes clear after reading Sects. 6.1 and 6.2 that the presented industrial and academic use cases do not provide fully grounded evidence of expected performance or user acceptance. Indeed, the industrial use case might have used an incomplete document collection for GAN and did not answer its question **Q3** using retrospective terminological difference analysis. As for the academic use case, the subject sample of 19 people is too small to ensure reliable conclusions. Indeed, these use cases are both works in progress. These projects will be continued to collect more evidence and experience. Currently, the use cases are presented as potentially good indications of the applicability of our research results in the contexts of these use cases for solving practically important and scientifically interesting problems.

[10] https://www.nltk.org/api/nltk.html.

6.4 Potential Business Use Scenarios in Scientific Publishing

The developed method and its software implementation are intended to help analyse an important property of representativeness of a domain-bounded document collection. It is implicitly assumed that the documents in such a collection are of good quality and reflect the understanding and interpretation of the focal subject domain by the domain knowledge stakeholders. Otherwise, the terminology extracted from the documents would not be appropriate for use in the development of the ontology of this subject domain. One industry that naturally owns the collections of documents having the above-mentioned properties is Scholarly Publishing; and the documents are scholarly papers collected around the subject domains related to research and development. Therefore, Scholarly Publishing was selected as a focal industrial sector for us to seek potential business applications of our research output.

Business applications of the presented method for computing saturated terminological footprints of text collections in the Scholarly Publishing industry are envisioned in developing software tools to instrument and support:

- Subject descriptor taxonomies refinement
- Journal portfolio management
- Publication venue recommendation for potential authors
- Recommendations of potential reviewers, editorial board or program committee members.

Subject descriptor taxonomies are known to be useful and used for structuring content in many digital libraries of scholarly publications and indexing services, such as ACM DL,[11] Microsoft Academic,[12] Scopus,[13] Web of Science.[14] It is also known that these taxonomies are mostly handcrafted and, therefore, are imprecise,[15] are incomplete, and do not evolve as quickly as the scholarly content. Using the updates in scholarly content for regular improvement of subject descriptor taxonomies might therefore be beneficial for digital library businesses.

Journal portfolio management, at program management level, often faces the problem of significant topical overlaps in different journal portfolios. For example, journals on operations research could be seen in Mathematics and Economics portfolios. For program managers, overlooking and coordinating different portfolios with an aim of improving their impacts and publication efficiency, an instrument for comparing the semantic "footprints" of different portfolios might be of considerable interest. For such a comparison, mapping the saturated term-sets extracted from

[11] https://dl.acm.org/.

[12] https://academic.microsoft.com/.

[13] https://www.scopus.com/

[14] https://webofknowledge.com/.

[15] For example, the topic of "saturation" is related to Chemistry, but not also to Computer Science or Qualitative Research.

different portfolio document collections or derived knowledge graphs might be an effective solution.

Publication venue recommendations are provided by many scholarly publishers via online venue recommendation services for their potential authors to select the best matching outlet for their submission. These services, for example, Springer Journal Suggester[16] or Elsevier JournalFinder,[17] are based on the natural language processing and analysis of a submitted text and finding matches to the scope subject descriptors of a journal portfolio. The results generated by these systems might be improved by refining the subject descriptors—as outlined above.

Recommendation of reviewers is one more interesting practical problem for editors or program chairs who have to match reliably the competencies and interests of the potential reviewers to the subject matter of submission. An informed reviewer suggestion might be offered based on the terminological comparison of the body of the publications by the candidates to the semantic footprint of a particular submission.

In the outlined business contexts, using our method of extracting terminological cores and corresponding saturated sets of terms might be a cost-effective, yet reliable way for instrumenting corresponding business activities.

6.5 Summary

This Chapter was focused on meeting our research objective **RO4** (c.f. Sect. 2.13.2). For that, the answers to research questions **RQ10** (Sects. 6.1 and 6.2) and **RQ11** (Sects. 6.3 and 6.4) were presented.

Section 6.1 reported on our experience of exploiting the developed software pipeline in an industrial use case project aimed at verifying the prognosis, by Gartner, on the trends of technology uptake with respect to the GAN technology. This use case demonstrated that our method and software pipeline could be used not only for extracting terminological core collections but also for constructively using the absence of saturation to analyse technology adoption trends.

Section 6.2 reported the use of our method and software in teaching for instrumenting the exploratory research of Master students with their aim to write the reviews of the related work for their Master theses. This use case demonstrated that the developed solution was user friendly and effective in these academic settings. It has also been well accepted and characterized as useful by the involved students.

Section 6.3 extended our use case experience by outlining the practical benefits of the use of our method and software for early industrial adopters. Our understanding of the limitations of the developed method in its current state was summarized. Finally, Sect. 6.4 presented our deliberations on the potential business scenarios and applications of our research output in the Scholarly Publishing industry.

[16] https://journalsuggester.springer.com/.

[17] https://journalfinder.elsevier.com/.

The reported use case experience, benefits and limitations summarization, and the outline of possible business application scenarios helped us understand the application and industrial sector scopes for the potential future transfers of our method and software to industry and academia.

References

1. Trends appear on the Gartner hype cycle for emerging technologies 2019. Gartner Inc. https://www.gartner.com/smarterwithgartner/5-trends-appear-on-the-gartner-hype-cycle-for-emerging-technologies-2019/. Accessed 14 Oct 2021
2. Cearley D, Jones N, Smith D, Burke B, Chandrasekaran A, Lu CK (2019) Top 10 strategic technology trends for 2020. Gartner Inc. ID: G00432920, 21 Oct 2019 https://www.gartner.com/en/doc/432920-top-10-strategic-technology-trends-for-2020. Accessed 14 Oct 2021
3. Dobrovolskyi H, Keberle N (2020) Obtaining the minimal terminologically saturated document set with controlled snowball sampling. In: Bollin A et al (eds) ICTERI 2020: Proceedings of the 16th international conference on ICT in education, research and industrial applications. Volume I: Main Conference. Kharkiv, CEUR-WS vol 2740. Oct 2020, p 87–101
4. Goodfellow I, Pouget-Abadie J, Mirza M, Xu B, Warde-Farley D, Ozair S, Courville A, Bengio Y (2014) Generative adversarial nets. In: Ghahramani Z, Welling M, Cortes C, Lawrence ND, Weinberger KQ (eds) Advances in Neural Information Processing Systems 27. Curran Associates, Inc., Boston, p 2672–2680
5. Gui J, Sun Z, Wen Y, Tao D, Ye J (2020) A review on generative adversarial networks: algorithms, theory, and applications. arXiv preprint, arXiv:2001.06937v1 [cs.LG]
6. Pan Z, Yu W, Yi X, Khan A, Yuan F, Zheng Y (2019) Recent progress on generative adversarial networks (GANs): a survey. IEEE Access 7:36322–36333. https://doi.org/10.1109/ACCESS.2019.2905015
7. Krizhevsky A, Sutskever I, Hinton GE (2017) IgeNet classification with deep convolutional neural networks. Comm ACM 60(6):84–90. https://doi.org/10.1145/3065386
8. Simonyan K, Zisserman A (2014) Very deep convolutional networks for large-scale image recognition. arXiv e-print: arXiv:1409.1556
9. Szegedy C, Liu W, Jia Y, Sermanet P, Reed S, Anguelov D, Erhan D, Vanhoucke V, Rabinovich A (2015) Going deeper with convolutions. In: CVPR 2015: 2015 IEEE conference on computer vision and pattern recognition, Boston, MA, pp 1–9. https://doi.org/10.1109/CVPR.2015.7298594
10. Ioffe S, Szegedy C (2015) Batch normalization: accelerating deep network training by reducing internal covariate shift. arXiv e-print: arXiv:1502.03167
11. Russakovsky O, Deng J, Su H et al (2015) ImageNet large scale visual recognition challenge. Int J Comput Vis 115:211–252. https://doi.org/10.1007/s11263-015-0816-y
12. Long J, Shelhamer E, Darrell T (2015) Fully convolutional networks for semantic segmentation. In: CVPR 2015: 2015 IEEE conference on computer vision and pattern recognition, Boston, MA, pp 3431–3440. https://doi.org/10.1109/CVPR.2015.7298965
13. Salvaris M, Dean D, Tok WH (2018) Generative adversarial networks. In: Deep Learning with Azure. Apress, Berkeley, CA, pp 131–160. https://doi.org/10.1007/978-1-4842-3679-6_8
14. Girshick R, Donahue J, Darrell T, Malik J (2014) Rich feature hierarchies for accurate object detection and semantic segmentation. In: CVPR 2014: 2014 IEEE conference on computer vision and pattern recognition. Columbus, OH, pp 580–587. https://doi.org/10.1109/CVPR.2014.81
15. Suthaharan S. (2016) Support Vector Machine. In: Machine learning models and algorithms for big data classification. Integrated series in information systems, vol 36. Springer, Boston, MA. https://doi.org/10.1007/978-1-4899-7641-3_9

16. Zeiler MD, Fergus R (2014) Visualizing and understanding convolutional networks. In: Fleet D, Pajdla T, Schiele B, Tuytelaars T (eds) ECCV 2014: 13th European conference on computer vision, Zurich. Lecture Notes in Computer Science, vol 8689. Springer, Cham, Sept 2014, pp 818–833. https://doi.org/10.1007/978-3-319-10590-1_53
17. Settles B (2009) Active learning literature survey. Computer sciences technical report 1648. University of Wisconsin–Madison. https://minds.wisconsin.edu/bitstream/handle/1793/60660/TR1648.pdf. Accessed 14 Oct 2021
18. Kosa V (2021) A method of experimental study of terminological saturation in document collections for knowledge elicitation. Ph.D. Dissertation, Zaporizhzhia National University (in Ukrainian). https://doi.org/10.13140/RG.2.2.35971.63525
19. Kosa V, Ermolayev V (2020) Toward a theoretical framework of terminological saturation for ontology learning from texts. In: MS-AMLV 2019: Proceedings of the 1st masters symposium on advances in data mining, machine learning, and computer vision. Lviv, CEUR-WS, vol 2566. Nov 2019, pp 40–51
20. Ermolayev V, Dobrovolskyi H, Kosa V, Yuschenko E (2020) Terminological analysis of the gaps between academic and industrial R&D in Generative Adversarial Networks. Technical Report BWT-SAGOIT-IT-TR-2020–1, GroupBWT, Ukraine, Dec 2020, 39 p

Chapter 7
Conclusions and Outlook

A method and its software implementation, that help draw correct and timely conclusions and recommendations based on the analysis of big available textual data, are being on demand and under intensive research and development today. One of the important aspects, that influences the validity of conclusions based on data analysis, is the representativeness of a data sample.

The assessment of data representativeness is particularly complex in ontology engineering. In this field, the data are the interpretations of a particular subject domain by the community of knowledge stakeholders in this domain. Representativeness in this context is the degree to which all these interpretations are taken into account in the specification of the requirements for the development of a grounded descriptive theory of the domain.

In this book, this problem was addressed by developing the method to solicit the domain interpretations of the majority of the knowledge stakeholders by extracting terminology from the sample of professional texts describing the domain. The developed method allowed automatically extracting a representative set of terms of minimal size and forming a terminological core sub-collection of documents, of minimal size, that contained all these terms. For that, the notion of terminological saturation as a successive approximation process leading to the terminological core sub-collection for an arbitrary subject domain was introduced.

Based on the extensive and systematic review of the related work, the research gaps were detected that needed to be bridged. Based on the analysis of these research gaps, an approach for saturated terminology extraction was proposed. With an aim to operationalize this approach in a method, the research questions of our study were put forward in Chap. 2. These research questions were grouped into four research objectives. Our research was organized and methodologically structured with an aim to meet these research objectives:

- **RO1 (formal framework)**: Develop a formal framework covering representations, formalisms, and proofs for all the elements required to formally present the successive approximation process for reaching saturation in a domain-bounded document collection. Prove the validity of this formal framework.

V. Kosa and V. Ermolayev, *Terminology Saturation*, Cognitive Science and Technology,
https://doi.org/10.1007/978-981-16-8630-6_7

- **RO2 (algorithmic suite)**: Based on the developed formal framework, develop the suite of algorithms that operationalize the elaborated method of successive approximation for saturated terminology extraction. Implementing these algorithms is the software suite.
- **RO3 (experimental evaluation and validation)**: Using the software suite, evaluate and validate the developed method for saturated terminology extraction on several synthetic and real document collections of scholarly publications belonging to varying subject domains.
- **RO4 (practical use)**: Deploy and try the validated method and software in industrial and academic use cases. Check if the contribution has potential for technology transfer. Find out what are the focal sectors for possible technology transfer and deployment.

In Sect. 7.1, the findings and results of the presented work are summarized. This summary shows that all the above-mentioned research objectives have been fully met in the study. The issues that yet remain unsolved are outlined in Sect. 7.2. This outline is followed by a concise presentation of the plans for future work.

7.1 Summary of Findings and Results

The developed formal framework, that met **RO1**, was presented in Chap. 3. The framework was a coherent formal representation of the successive approximation method for measuring terminological saturation in text collections.

It dealt with the three important ingredients of the method: (i) the function for measuring terminological difference (thd) and its formal properties; (ii) the sufficient existence conditions for terminological saturation; and (iii) the formal approach for optimizing the computation in the process of extracting terms. To develop these ingredients, three respective research hypotheses, **H3.1** (*thd is a metric function*), **H3.2** (*existence of terminological saturation*), and **H3.3** (*Merged Partial C-values are equal to conventional C-Values*), were formulated and further tested as valid in Chap. 3.

H.3.1 was formally tested in Sects. 3.3 and 3.4. The function for measuring terminological difference (*thd*) between the bags of terms was formally defined in Sect. 3.3 as a kind of the Manhattan distance function. The definition of thd indirectly used the definition of the individual term significance threshold (*eps*) that was introduced in Sect. 3.1. The metric properties of the *thd* function were formally validated by the proofs of Lemma 3.1 and Theorem 3.2 in Sect. 3.4.

H3.2 was formally validated in Sect. 3.5 by the proof of the existence Theorem 3.3. This result, however, did not allow computing the predictions of saturation points in arbitrary document collections, based on a few initial iterations.

H3.3 was formally tested in Sect. 3.6. For testing the hypothesis, the notions of a partition of a document collection, a partial C-Value, and a merged partial C-Value were introduced. Further, the proofs of Lemma 3.4 and, partially, of Theorem 3.5

formally validated that conventional C-Values were equivalent to merged partial C-values up to the validity of the $h1$ hypothesis about the approximate equality in $\sum_{i=1}^{n}\left(\frac{1}{P(T_i^s)}\right) \approx \frac{1}{P(T^s)}$. This hypothesis ($h1$) was proven experimentally in Sect. 5.7, which finalized the validity check of **H3.3**. As a useful consequence of validating **H3.3**, the Corollary 3.6 to Theorem 3.5 was proven. It stated that merged partial C-values did not depend on the size of a partition dataset. Therefore, the size of a partition dataset might be reasonably chosen based on the specifics of the problem and available hardware resources (e.g. random access memory). This result, in fact, removed the constraints of the software implementations of the C-Value method and allowed our MPCV method to scale efficiently up to the realistic industrial sizes of document collections.

The developed algorithmic suite for terminological saturation measurement, that met **RO2**, was presented in Chap. 4. The suite implemented our formal framework for terminological saturation measurement (Chap. 3) as a computational method. In particular, the algorithms developed in this chapter and implemented in the software suite (Sect. 4.8) were used for the experimental evaluation of the developed computational method in Chap. 5 and in the use cases presented in Chap. 6. Furthermore, the software suite was made publicly available for research purposes in order to facilitate the reproducibility of our results.

The algorithmic suite was developed for automating or partially automating the elaborated workflow for terminological saturation measurement (c.f. Sect. 4.1, Fig. 4.1) at the task level of detail. The tasks were grouped in phases. For the purpose of the development of the algorithmic suite, the workflow was further detailed in the computation flow (c.f. Sect. 4.1, Fig. 4.3). Based on this computation flow, which specified the modules (functions), data flow and transformation, iterations, and termination criteria, the algorithms for all the modules, that require the use of software, were elaborated in Sects. 4.2–4.7. Section 4.8 reported about the implementation of the algorithmic suite in the software. Hence, the software instruments, required to receive the full answers to the research questions **RQ1**, **4**, **5**, and **7** (c.f. Sec. 2.13, Table 2.5), were obtained.

Yet furthermore, Sect. 4.6 elaborated the formal and algorithmic approach to answering our research question **RQ8** about the utility of terms grouping for terminological saturation measurement. Based on the research hypothesis **H4.1**, put forward in Sect. 4.6, four relevant string similarity measures were selected, term similarity thresholds for the selected string similarity measures were devised, and the necessary algorithms for similarity measurement, terms grouping, and terminological difference measurement were developed.

Chapter 5 focused on meeting our research objective **RO3**. For that, our series of experiments were set up, performed, their results discussed, and several recommendations drawn.

For the experimental set-up (Sect. 5.2), the objectives were formulated and experiments were planned to meet these objectives. Experimental objectives were formulated in Sect. 5.1 in a way to fully cover the evaluation and validation facet of the research questions **RQ9**, **3**, **1**, **8**, and **7** (c.f. Table 2.5). As the parts of the plan: (i) the experimental workflow was developed; (ii) the required data was collected; (iii) the measurable aspects and measures for assessing the results were specified; and (iv) the environment for experimental runs was configured.

In Sect. 5.3, **RQ9** about the correctness of the developed method and algorithmic suite in the boundary cases was positively answered. For that, the experiments using the synthetic collections, 1DOC and RAW, were performed. 1DOC and RAW had been designed to represent the data for the boundary cases: quick and steady saturation and no saturation. The results of the experiments demonstrated that the method and its algorithmic and software implementations produced expected results therefore were correct.

In Sect. 5.4, **RQ3** about the choice of the most appropriate ATE software tool for terminological saturation measurement was answered. As a result, UPM Term Extractor software was selected based on the cross-evaluation against NaCTeM TerMine. UPM Term Extractor was further used in all the computations as a part of the baseline computation pipeline.

Section 5.5 answered **RQ1** about the impacts of different possible document orderings on terminological saturation measurement. Based on cross-evaluating five different orderings against the measurable aspects (Table 5.2), the descending citation frequency (**dcf**) ordering was recommended as the most balanced ordering across different aspects and subject domains.

Section 5.6 evaluated if the use of similar terms grouping had positive impact on terminological saturation (**RQ8**). In this evaluation, the refined computational pipeline, with terms grouping and terms similarity measurement, was applied to three different real document collections in different subject domains. In the experiments, the methods for term similarity measurements and corresponding terms similarity thresholds were varied in order to find the most productive configuration regarding the quality of outputs and computational efficiency. As a result, two best-ranked configurations were recommended for use.

Finally, Sect. 5.7 answered **RQ7** about the validity, efficiency, and scalability of the optimized computational pipeline. The hypothesis $h1$ of Theorem 3.5 was proven experimentally, which finalized the proof of this theorem. It was also found out that the optimized pipeline was computationally efficient, straightforwardly parallelizable, and scaled sufficiently well to waive the inherent memory constraint of the State-of-the-Art ATE implementations. Hence, it was recommended that the optimized pipeline could be used for processing industrial size document collections.

Thus, our research objective **RQ3** was met.

Chapter 6 focused on meeting our research objective **RO4**. For that our answers to research questions **RQ10** (Sects. 6.1 and 6.2) and **RQ11** (Sects. 6.3 and 6.4) were given.

Section 6.1 reported on the demonstration of the use of the developed software pipeline in an industrial project aimed at verifying the prognosis, by Gartner, on the trends of technology uptake for the example of the GAN technology. This use case indicated that our method and software pipeline could be used not only for extracting terminological core collections, but also to analyse technology adoption trends based on the presence or absence of terminological saturation.

Section 6.2 demonstrated the use of our method and software in teaching—for instrumenting the exploratory literature research of Master students with their aim to write the reviews of the related work for their Master theses. This use case demonstrated that the developed solution was effective in these academic settings. It has also been accepted and characterized as useful by some of the involved students.

Section 6.3 generalized our practical experience by outlining the potential benefits and current limitations of our method and software for early industrial adopters. Section 6.4 presented our view on the potential business scenarios and applications in the scholarly publishing industry.

The reported use case experience, summarization of practical benefits and limitations, and the outline of possible business application scenarios helped us understand the application and sector scopes for the potential future transfers of our method and software to industry and academia. Thus, our research objective **RQ4** has been met.

The research objectives put forward and finally met in this work were in fact targeted at improving the representativeness, efficiency, and effectiveness of automated terminology extraction from the collections of professional high-quality textual documents, like scholarly publications. The summary of our research output and outline of what was the measured improvement along the dimensions of representativeness, efficiency, and effectiveness are as follows.

A formal framework for the method of terminological saturation detection and measurement was developed. The framework comprised:

- The formal representation of the process of successive approximation to a terminological core sub-collection
- The metric of terminological difference between the bags of significant retained terms
- The specification of the sufficient conditions for the existence of terminological saturation
- The optimized processing pipeline based on the use of the partition of a document collection.

The developed formal framework was operationalized in a suite of algorithms. The algorithms were implemented in the software modules that were aggregated in the processing pipelines (baseline and optimized) for the detection and measurement of terminological saturation.

The implemented processing pipelines were systematically evaluated in a series of computation experiments. Several recommendations were elaborated and given, based on the results of these experiments. Elaborated recommendations allowed offering optimally balanced configurations of the processing pipelines.

The experiments also proved that the developed method was correct, efficient, effective, and domain neutral.

The **efficiency** of the developed and implemented method was proven as follows.

(i) **Compactness**. If terminological saturation existed in the processed collection of documents, the method yielded a substantially more compact terminological core sub-collection, compared to the entire collection. The saturated bags of

Table 7.1 Compactness of terminological core sub-collections and saturated bags of retained significant terms

Collection	Number of documents		Number (Percent) of retained significant terms		
	Entire collection	Terminological core	Entire collection	Terminological core	With terms grouping
DMKD	300	80 (26.7%)	313,506	3399 (1.08%)	2135 (0.68%)
TIME	437	180 (41.2%)	315,474	5493 (1.74%)	3456 (1.10%)
DAC cleaned	506	220 (43.5%)	518,765	16,703 (3.22%)	7496 (1.44%)
KM	7500	500 (6.6%)	4035 760	13579 (0.34%)	–

retained significant terms were also substantially more compact than the bag of terms extracted by the same ATE method from the entire document collection. It was proven in our evaluation experiments that the compactness of the outputs was observed independently from the subject domain to which the collection belonged. The illustration of the compactness of the method output for the real document collections used in our experiments is given in Table 7.1. It was also experimentally proven that the saturated sets of grouped terms were even more compact, irrespectively to the domain of the collection. However, this gain came with substantially (times) higher computational expense.

(ii) **Time of computation**. The developed optimized pipeline took significantly less time than the baseline pipeline for processing the same input text datasets. It was achieved due to the use of partition datasets and the trie data structure in the statistical part of the implementation of the MPCV method. As it was demonstrated in Sect. 5.7.2 reporting the experiment that used the DMKD collection, the optimized pipeline spent practically the same time per each iteration (32–51 s). The baseline implementation spent from 30 s for the 1st iteration to 460 s for the 15th iteration.

The **effectiveness** of the developed method was proven as follows.

(iii) **Representativeness**. In case terminological saturation existed for a document collection, the method extracted a terminological core sub-collection that was guaranteed to be representative for the respective subject domain.

(iv) **Scalability**. Due to the use of partition datasets in the optimized processing pipeline, the method was not constrained by the size of an input collection. Due to practically constant time for the processing of a partition dataset and clear possibility of parallelizing the computation, the optimized pipeline scaled well with the growth of the input document collection to the volumes typical for industry. Scalability was demonstrated in our experiment with the KM collection that had typical industrial size (over 7,400 full-text journal articles).

The potential practical utility of the research output was demonstrated in our industrial and academic use cases. The discussion of this initial experience of practical use was complemented by the analysis of the benefits and limitations of the developed solution. Our initial thoughts about the potential business scenarios for the use of our method in the Scholarly Publishing industry were finally offered.

7.2 Open Research Issues and Future Work

The future work on further developing and refining the results presented in this book is envisioned and planned along the following two pillars:

- Method and technology refinement and extension
- The development of the additional use cases and technology transfer.

With respect to the further development of the method and technology for terminological saturation measurement and analysis, several open R&D issues might be summarized, that require future elaboration beyond the work presented in this book.

Prediction of terminological saturation. As acknowledged in Sect. 3.7, our proof of Theorem 3.3 on the sufficient conditions for the existence of terminological saturation does not yet allow us to constructively employ these conditions for predicting terminological saturation based on the subset of initial measurement iterations. The extension of our formal framework in this direction is planned for future work along the analysis of the statistical properties of term distributions in domain-bounded document corpora.

Cross-language document collections. One of the current constraints of our method and its software implementation is that it is applicable only to the documents written in English. It might be useful, however, to broaden the application scope of the solution by covering languages other than English and the collections containing documents written in several languages. As the method is language neutral except the part for terms extraction (c.f. Sect. 4.8), the solution might be extended to the other languages by adding term extractor modules for these languages, or cross-language term extractors, to the software suite. This extension is also planned for future work.

Developing a Platform-as-a-Service. Currently, the implemented instrumental software suite is publicly available as a research prototype. In our use cases (c.f. Sects. 6.1 and 6.2), it has been deployed ad-hoc for these particular uses. The next step is to develop and deploy an instrumental software platform, which can be seamlessly used by those who require terminological saturation analysis in their research or business. The development will be done in the future and in cooperation with an industrial partner.

As for the technology transfer activities, these will be planned based on the identification of potential business application scenarios, as presented in Sect. 6.4.

The manufacturer's authorised representative in the EU is Springer Nature Customer Service Centre GmbH, Europaplatz 3, 69115 Heidelberg, Germany. If you have any concerns regarding our products, please contact ProductSafety@springernature.com

Printed and bound by CPI Group (UK) Ltd, Croydon, CR0 4YY
30/09/2026
02225333-0001